About Island Press

Island Press, a nonprofit organization, publishes, markets and distributes the most advanced thinking on the conservation of our natural resources—books about soil, land, water, forests, wildlife, and hazardous and toxic wastes. These books are practical tools used by public officials, business and industry leaders, natural resource managers, and concerned citizens working to solve both local and global resource problems.

Founded in 1978, Island Press reorganized in 1984 to meet the increasing demand for substantive books on all resource-related issues. Island Press publishes and distributes under its own imprint and offers these services to other nonprofit organizations.

Support for Island Press is provided by Apple Computer Inc., Geraldine R. Dodge Foundation, The Energy Foundation, The Charles Engelhard Foundation, The Ford Foundation, Glen Eagles Foundation, The George Gund Foundation, William and Flora Hewlett Foundation, The Joyce Foundation, The John D. and Catherine T. MacArthur Foundation, The Andrew W. Mellon Foundation, The Joyce Mertz-Gilmore Foundation, The New-Land Foundation, The J. N. Pew Jr. Charitable Trust, Alida Rockefeller, The Rockefeller Brothers Fund, The Rockefeller Foundation, The Florence and John Schumann Foundation, The Tides Foundation, and individual donors.

About The Freedom Forum Media Studies Center

The Freedom Forum Media Studies Center, the chief operating program of The Freedom Forum, is the nation's first institute for the advanced study of mass communication and technological change.

Through a variety of programs, the Center seeks to enhance media professionalism, foster greater public understanding of how the media work, strengthen journalism education, and examine the effects on society of mass communication and communications technology. Among Center programs are residential fellowships, seminars, conferences, publications, and technology studies, including a state-of-the-art technology laboratory. The Freedom Forum Media Studies Center, Columbia University, 2950 Broadway, New York, New York 10027-7004, (212) 280-8392.

Media and the Environment

MEDIA
and the
Environment

Edited by

Craig L. LaMay

Everette E. Dennis

The Freedom Forum Media Studies Center

ISLAND PRESS

Washington, D.C. ■ *Covelo, California*

Originally published as a shorter work, "Covering the Environment,"
Gannett Center Journal, Vol. 4, No. 3, Summer 1990.

Changing the World Through the Informationsphere reprinted by permission
of Sterling Lord Literistic, Inc. Copyright © 1990 by Dana Meadows

Library of Congress Cataloging-in-Publication Data
LaMay, Craig L.
    Media and the environment / Craig L. LaMay and Everette E.
Dennis.
        p.   cm.
    Includes bibliographical references and index.
    ISBN 1-55963-131-7 (alk. paper). — ISBN 1-55963-130-9
(alk. paper : pbk.)
    1. Environmental protection in the press—United States.   2.
Environmental health—United States—Reporting.   I. LaMay,
Craig L.   II. Title.
PN4888.E65L36   1992
070.4'493637'00973—dc20   91-23442
                                                              CIP

Printed on recycled, acid-free paper

Manufactured in the United States of America
10   9   8   7   6   5   4   3   2   1

Contents

Foreword   *xi*
Bill McKibben

Preface   *xiii*

INTRODUCTION

Survival Alliances   3
John Maxwell Hamilton

SCIENCE, SOCIETY AND THE MEDIA

Two Decades of the Environmental Beat   *17*
Sharon M. Friedman

Of Science and Superstition: The Media and Biopolitics   *29*
John Burnham

An Odd Assortment of Allies:
American Environmentalism in the 1990s   *43*
Robert Gottlieb

In Context: Environmentalism in the System of News   *55*
Everette E. Dennis

COVERING THE ENVIRONMENT AS IF IT MATTERED

Changing the World Through the Informationsphere   *67*
Donella H. Meadows

Network Earth: Advocacy, Journalism and the Environment 81
Teya Ryan

The Traditionalist's Tools (And a Fistful of New Ones) 91
Jim Detjen

Heat and Light: The Advocacy-Objectivity Debate 103
Craig L. LaMay

Think Locally, Act Locally 115
William J. Coughlin

Media, Minorities and the Group of Ten 125
Gerry Stover

Who Speaks for the Land? 135
Dana Jackson

ECONOMICS AND ENVIRONMENTAL POLICY-MAKING

Boundless Bull 149
Herman E. Daly

Greens and Greenbacks 157
Emily T. Smith

A Market for Change 171
Timothy E. Wirth

Steering by the Stars 179
Albert Gore Jr.

Old Growth and the Media: A Lawmaker's Perspective 185
Mark O. Hatfield

THINKING GLOBALLY

A Green World After the Cold War? 195
John McCormick

Hungarian Greens Were Blue 205
Judit Vesarhelyi

Winds From the West   *217*
Aditia Man Shrestha

                           BOOK REVIEW

Books (Not Thneeds) Are What Everyone Needs   *225*
Robert Cahn

Annotated Bibliography   *246*
Index   *257*

Foreword

The environment will always be a hard story for journalists to cover. Our profession has, for the most part, described the conflicts between people—wars and crimes and politics and poverty and basketball. In some ways, of course, the story of our destruction of the environment continues this pattern—it is essential to report on the conflicts between rich and poor nations, between corporations and municipalities and people who live on the edge of waste dumps, between scientists with different opinions of our problems and different opinions about how to solve them.

But we stand at an interesting moment in history, when the chief conflict in the world may soon involve people and nature. If the researchers are correct about the greenhouse effect, for instance, then no story will outweigh our attempts to cope with 5-degree increases in temperature in the next six decades. If millions of species are in danger of being wiped out, we must ask ourselves hard questions about who we are as a species, and how far our rights extend. To cover this story accurately will involve, I think, enlarging our conception of our jobs. Topics long left to theologians, to nature writers, to philosophers must also become the province of women and men with reporter's notebooks and minicams.

This book, with its many fine pieces, is one indication we've started this task. May it spur others in the same direction.

Even the most remarkable coverage of environmental crises may not carry the day—the forces tending toward the destruction of ecosystems are powerful, deeply ingrained and in some cases irreversible. But there is no question that this is the most exciting, mind-stretching and important journalism that will be done in the years to come, and no question that we must do it well to have any chance at all of preserving the world as we have known it.

Bill McKibben
Author of *The End of Nature*

Preface

American thinking about the environment dates at least to Thomas Jefferson, who thought that democratic societies should preserve regions of scenic beauty for all their citizens to enjoy. In the middle and late 19th century, concern for the land and its wildlife found new voices in such men as Henry David Thoreau and John Muir, whose reverence for nature led them to advocate preserving it in its wildness, and George Perkins Marsh and Gifford Pinchot, who saw utility in nature as well as beauty and so advocated principles of resource management, or conservation. At the turn of the century, Muir and Pinchot competed intensely, sometimes bitterly, for the ear of President Theodore Roosevelt, whose record as an "environmental president" exceeds that of perhaps any other chief executive.

Today we have another "environmental president," and as a topic for media attention the environment has probably never fared better, as the essays in this book attest. But journalists and scholars alike have doubts about the media's rekindled interest. Some wonder whether it is simply an attempt to tap what poll data show is intense (and possibly fickle) audience concern about these issues, particularly as they affect public health and the quality of life. Will continued uncertainty about oil prices and availability scuttle the public's concern for the environment, and with it media coverage, in the same way that the Arab oil embargo did in the 1970s?

The resounding answer contained in these pages is no: As journalists have sought to find their way through the complex and far-flung web of scientific, political, economic and philosophical concerns that shape environmental issues, they have made too many active connections between the degradation of the Earth and the behavior of its most enterprising inhabitant. In short, they say, the course and consequences of human activity have already determined the environment's place as a leading public issue for the 1990s and beyond. The questions that remain concern how to cover it. Many of them are profoundly difficult, and some challenge established tenets of American journalism.

IN DEVELOPING THIS book, we sought intellectual guidance from many people whose contributions to environmental research and writing have made them leaders in the field. Time after time those conversations would turn to a single problem: how to reconcile the emerging media picture of humanity growing into self-extermination with coverage of other topics, from public health and private enterprise to civil rights and national security. At another level, concern for this kind of cognitive dissonance extended to the media's important role as an advertiser, as a virtual conduit for consumption.

Some rightly point out that many of these connections are still quite new to reporters who ply the environmental beat, to say nothing of their editors and their audience. Several leading media organizations have made public commitments to addressing these problems, installing environmental reporters in such once unlikely places as the business staff.

Others have gone a step further and proclaimed themselves environmental advocates. Perhaps the most well-known of these proclamations came from Time senior editor Charles Alexander at a 1989 Smithsonian Institution conference on environmental reporting, an announcement that, coming shortly after Time's celebrated "Planet of the Year" issue, caused enough of a stir that Alexander later sought to explain his position more carefully. He has not been alone. The idea that the environment is a story in need of a new kind of reporting, if not outright advocacy, is one that most of the authors in this book have had to reckon with.

For some of them, however, advocacy is secondary to their belief that we must change the focus of the debate about what in this world has value—whether economic, social or moral—and the questions they raise are for the most part still foreign to media coverage of the environment: Is a nation's consumption a meaningful measure of its economic health? Can growth itself be destructive of the public good, logically and even morally suspect? Is poverty one of the most important global environmental issues? Has the information system we depend on to address these issues broken down?

Some in the media are already beginning to ask such questions, but for the most part coverage of the environment has yet to break free from the traditional demands of "news" or existing paradigms of economic and social thinking. It is worth noting, for example, that much of today's environmental coverage—if not most of it—is about pollution rather than conservation. Stories about preservation—about the values of nature for nature's sake—are few and generally poorly done. They are usually of the variety that suggest a simple if painful choice between an owl and an industry, to take a recent example, and do little to examine either the nature of the industry or the *nature* of what is to be preserved. Stories about pollution simply fit more readily into the requirements of news than do articles about preservation, which tend to wind up as "magazine" or feature pieces.

Even many of the issues that do get front-page treatment, that are routinely touted as "global," often get snared in parochial thinking. When, for example, the Western economic powers met in Houston in 1990, they postponed action on limiting their own contribution to the buildup of greenhouse gases but vowed to bring pressure on Third World countries to reduce their rate of deforestation—a resolution that brought little if any editorial comment in American media.

Other issues with obvious global implications almost never escape the parameters of provincialism. Coverage of birth control, for instance, is usually couched in the context of American constitutional law, despite the fact that the worldwide pressure of populations on natural resources is an increasingly critical environmental and social-justice issue. Similarly, pesticides that have been banned or restricted in the United States for nearly 20 years are still exported for agricultural use in other countries—often to find their way back into the U.S. food supply as imports. Only recently has Congress—or the media—given this problem much attention, and even then much of the press attention has missed or ignored larger issues. On-going negotiations on the General Agreements on Trade and Tariffs, for example, stand to make existing U.S. and state environmental safety and health protection laws virtually impotent in the face of much-less-demanding international standards, but until recently that aspect of the GATT talks has received almost no coverage at all.

BEGINNING WITH John Maxwell Hamilton's introductory essay, "Survival Alliances," this book speaks to the many different and divergent currents that together comprise "the environment story." Hamilton, a former foreign correspondent and senior World Bank official, chooses a single environmental issue—the destruction of tropical biodiversity—and finds it virtually impossible to explain except in the context of other economic, political and philosophical issues, each of which poses its own unhappy dilemmas. Journeying to "biological superpower" Costa Rica, Hamilton describes a country whose rich variety of plant and animal life is essential to agriculture and medicine in the developed world but that is also pressured by population growth, poverty and foreign debt. Day by day, hour by hour, the nation plunders its promise just to survive. Should Americans give high priority to guarding Costa Rica's genetic diversity, Hamilton asks, or should they emphasize its repayment of commercial loans to U.S. banks? In light of our own environmental mismanagement, is it surprising that tropical Third World countries hear in our cries to save

their forests a kind of hypocritical neo-imperialism? Even if they
do, and even if we were to admit it, can anyone anywhere afford to
ignore the destruction of vital world resources? "The environment is a prime candidate to replace East-West confrontation as
our No. 1 national security consideration," Hamilton writes,
"only the issues will be more, not less, complicated. . . . Journalists have . . . a lot of explaining to do."

Following the introduction, four essays examine environmentalism as a topic for media coverage and public debate. Journalism
educator Sharon Friedman of Lehigh University looks at the development of the issue in the news media and argues that while
coverage has boomed and diversified, it still suffers from the fundamental problems that afflicted it in 1970, when we first celebrated Earth Day. John Burnham, a historian from Ohio State
University, also finds fault with media coverage of the environment, commenting that in the interplay of science, journalism and
public policy, environmentalism has been largely stripped of its
educational and moral values and rendered impotent except as a
political issue. Policy analyst Robert Gottlieb of the University of
California at Los Angeles wonders whether the news media
haven't missed much of the "process" part of the story by focusing
on the so-called Gang of Ten—the big, national groups that get to
speak in the name of the environment. Many of the most important victories for environmentalism in the past decade have been
won at the grass roots level, Gottlieb says, a trend that is likely to
grow stronger in the 1990s. Closing the section, educator and
media critic Everette E. Dennis, a co-editor of this book, argues
that journalists and the public alike often overestimate what the
American system of news and information is capable of and have
little understanding of the decision-making process that guides it.
Environmental issues, Dennis says, may be important, but in the
news game they are simply another competing interest.

The next section of the book, "Covering the Environment as if
It Mattered," forms the heart of the book. Syndicated columnist
Donella Meadows leads the section by urging journalists to dis-

cover new paradigms for covering the environment, for shaking themselves—and the public—out of the cognitive dissonance that allows a story about global warming to run next to a happy report about a rise in gross national product. Turner Broadcasting's Teya Ryan picks up a strand of this theme in her defense of environmental advocacy, and *Philadelphia Inquirer* reporter Jim Detjen, president of the Society of Environmental Journalists, answers that traditional reporting methods—coupled with new technologies and backed by editorial commitment—can be just as effective as advocacy and, in the long run, more credible. *Gannett Center Journal* editor Craig LaMay, the other co-editor of this book, follows with an essay arguing that the advocacy-objectivity debate is more semantics than substance, but that underlying the advocacy movement are pointed criticisms of the values that shape news generally and which are worth serious attention.

The difficult choices that come with environmental-economic trade-offs don't get any easier when the focus of your reporting is your next-door neighbor, as editor William Coughlin found out when his newspaper undertook an investigation of the water supply in the small, rural town of Washington, North Carolina. The resulting stories won a Pulitzer for Coughlin's *Washington Daily News*, but they didn't win him many friends. Following Coughlin is an essay by Gerry Stover, formerly of the Environmental Consortium for Minority Outreach, who reminds us that "there can be no safe environment without a just environment" and asks why the Gang of Ten and the media have failed to notice the environmental problems that are of particular consequence to poor and minority communities. Agriculture writer Dana Jackson of The Land Institute then looks at news coverage of land, soil and water. Historically one of the most productive and commanding areas of environmental news, agriculture is nonetheless poorly understood, Jackson writes, and nothing better illustrates the point than the media's failure to cover the ongoing GATT talks and their inability to understand what's at stake in them.

In a special section on economics and policy-making, U.S.

Senators Timothy Wirth of Colorado and Albert Gore Jr. of Tennessee each respond to a series of questions about media coverage of the environment, and Senator Mark Hatfield of Oregon offers a critical view of the media's coverage of the Pacific Northwest's old-growth forests, jobs and wildlife.

Leading the section is a masterful essay by World Bank economist Herman Daly, "Boundless Bull," which takes orthodox economics to task for its irrational devotion to growth as a measure of the common good and questions whether journalists will ever develop the moral outrage or clear thinking to challenge it. Following him, Emily T. Smith, science editor for *Business Week*, follows with a look at the environment as a leading business story in the 1990s.

The truly global nature of environmentalism is often lost on Americans, who tend to see the movement as another of their democracy's gifts to the world. Nothing could be further from the truth, of course, and so this book offers three international views of environmental change and the role of the media in it. John McCormick, a scholar in the field of international environmental policy, looks at the rise of Green politics in Europe, while Hungarian journalist Judit Vesarhelyi describes how a fledgling "Blue" movement employed underground newsletters and Western media to rally virtually an entire population against the damming of the Danube. Writing from Kathmandu, journalist Aditia Man Shrestha looks at the peculiar problems that attend covering the environment in a region beset by nationalism and where reliable information is hard to come by.

Finally, a book review of classic texts looks at the especially profound impact that that medium has had on American environmental thought. Written by Pulitzer Prize-winning journalist Robert Cahn, the essay is a literary excursion through the environmental ethic and a reminder that it was shaped by men and women of exceptional vision and courage.

In 1962, biologist Rachel Carson established in the public mind the simple principle that environmental degradation anywhere has unforeseeable consequences elsewhere, that we are all part of the environment, and that we cannot poison the Earth, its waterways or its air without eventually poisoning ourselves. This seemingly self-evident truth, Carson realized, ran smack into our everlasting and arrogant certainty that man can control nature, that we can survive our own ingenuity. This belief has been rooted in Western culture since the time of Francis Bacon, and the unfortunate price of its optimism is that it dismisses virtually any evidence of its failure and leaves open the possibility that we might one day, quite unknowingly, cross a threshold into a socioecological abyss from which we cannot escape.

In its eloquence and passion, *Silent Spring* brought us this message nearly 30 years ago. Carson was one of those rarest of talents—a brilliant scientist who was also a great writer, a woman who could see science in the context of what was just and moral. Without others like her, society will depend on reporters and editors to perform the tasks of research, synthesis and communication. If media coverage of the environment is to have any significance for the public or for policy-makers, it will have to do more than offer data. Writing on the interplay of science, politics and ethics, political scientist Lynton Caldwell has commented that "the mere possession of scientific knowledge holds no promise of its use in discovering the true needs of men or in serving the public happiness or welfare. There is need for more knowledge—but even greater need for more understanding. Development of valid and coherent concepts of man in nature requires an interrelating and a synthesizing of knowledge. It is a task of interpretive leadership." That is the goal to which this book is dedicated.

C. L. L.

E. E. D.

INTRODUCTION

The Earthly Paradise: John Parkinson, London, 1629
New York Public Library

JOHN MAXWELL HAMILTON

Survival Alliances

———

Flashlight in hand, I uneasily traversed the chain-and-board bridge swaying high over the river and warily made my way along the dark, forest-enclosed trail to my screened-in quarters.

This was the La Selva Biological Station in northeast Costa Rica. During the day I had seen just about every creature imaginable, I thought: a snake (rather close to my head) that mimicked a tree branch; dragonflies propelling themselves by two pair of improbably iridescent blue wings; 3-foot-long iguanas perched on tree limbs; and bright green lizards, called Jesucristos, scooting across the river's surface. At dinner, the small group of visiting U.S. scientists seemed yet another exotic species as, here in the steamy jungle, they ardently discussed their specialized work and difficulties getting published in scholarly journals back home. When a large insect landed on my shoulder and bit, one rotund ant specialist regarded it clinically and then, begrudgingly, swatted it.

This was a trip to see close-up the diverse plants and animals residing in developing countries. Writing a book on global interdependence, I was devoting one chapter to industrialized countries' dependence on these fragile biological cornucopia. As I walked along the dark path, the tropical world seemed inhospita-

ble, utterly unfamiliar, and exceedingly difficult to explain to readers accustomed to air conditioning and manicured suburban
lawns.

Later, after completing two trips to Costa Rica, I would see a
metaphor for the entire experience in those moments at the beginning of my research. The environment story has become one
of the biggest of our time. In 1988 *Time* magazine strayed from its
tradition of naming a person of the year and declared "Endangered Earth" the "Planet of the Year." "ABC Nightly News" ranks
the environment one of the five prime topics on its "American
Agenda" feature; it has two environmental reporters. This is as it
should be. Degradation of air, water and soil are life-and-death
issues. They demand media attention. Yet data are just beginning
to come in on ozone depletion as a result of the use of chlorofluorocarbons (CFCs) or on the Earth's warming due to the accumulation of greenhouse gas in the atmosphere. Beyond that, a
thicket of uncharted global political, economic and social considerations surround environmental problems. Journalists have every
reason to trace these complicated paths through environmental
interdependence and every reason to feel cut loose from their
moorings as they do.

I chose the subject of biological diversity precisely because it is
one of the less-well-appreciated global environmental stories.
That this is so should come as no surprise. The classification,
preservation and use of different species of plants and animals
pushes at the frontiers of science. Research on microscopic genes
from remote countries, however, does not produce the noisy and
telegenic excitement of a Hubble satellite launch. And explaining
the process by which these exotic plants and animals enhance our
lives often leaves the experts tongue-tied.

"Why should we care?" I asked an eminent scientist, hoping for
a vivid explanation of the importance of Costa Rica's biological
wealth to people in the United States.

"Because," he said, "if you live in Des Moines and care about
New York City, you should care about Costa Rica."

The place to start filling in this elusive equation is Costa Rica, sitting only 10 degrees off the equator. The tropics, made up largely of developing countries, are blessed with the world's greatest variety of plant and animal species, thanks to their abundance of the ingredients of life: sunlight and water. Overall the tropics hold well over 75 percent of all plant and animal species, although they occupy only 42 percent of the Earth's land area. Costa Rica, at the junction of two continents and having both high mountains and coastlines, is a "biological superpower," as one of its ministers puts it. By some estimates it has more diversity per square foot than any other place on Earth.

"Even in absolute numbers, you find that Costa Rica has more species of birds than North America, including northern Mexico, the United States, Canada and Alaska," says Rodrigo Gámez, director of Costa Rica's National Institute of Biology. "The La Selva Biological Station has almost two times the number of plants and animals of the state of California. And we're just talking about 1,300 hectares."

The fabulous array of plant and animal life is more than a scientific curiosity: Citizens of industrialized countries are dependent on tropical plants and animals. One common example lies in medicine. A cure for childhood leukemia came from the rosy periwinkle in Madagascar, and scientists speculate that a cure for AIDS may come from the tropics. Another example, which especially intrigued me, is the dependence of the American farmer, whose crops are powered by foreign genes just as surely as our industry is powered by foreign oil.

North Americans think of their nation as a land of agricultural abundance, but virtually all of its major cash crops originated elsewhere. Paradoxically, this "elsewhere" is largely in so-called less-developed countries. Soybeans come from China, wheat from the Fertile Crescent, and corn from Central America. While these crops have long been North Americanized, their points of origin remain important to U.S. farmers. These genetic homelands contain the widest variety of different species, numbering in the tens of thousands for crops like corn and wheat and often growing in the wild. Each species has unique genes, molecules

that bear an inherited code that determines specialized characteristics. The genes are essential to the creation of improved hybrid varieties of crops that survive drought, resist disease or bear
more fruit.

Anyone who thinks that reducing our dependency on foreign
oil is difficult should look at our dependency on foreign germplasm, as these gene resources are called. Farmers constantly
need new hybrids—and hence new genes—if they are to stay
ahead of pests, which themselves adapt to the hearty feasts improved crops offer. In addition, no variety, however well it per-
forms, is immune to all diseases. When farmers rely on only a
small range of varieties and the right disease comes along, the
consequences can be catastrophic. That was at the root of Ire-
land's 19th-century famine; the Irish used only two samples of
potatoes from the Andes, where the crop originated.

The uses of individual genes become more sophisticated all the
time. Scientists are no longer limited to combining genes from
different species of corn, for example. They are developing techniques to pluck an individual gene out of one plant and put it into
an entirely different kind of plant, say to "inoculate" food crops
against a particular virus. Recent experiments have even involved
putting animal genes into plants.

None of this, however, happens neatly or quickly. Decades
passed before scientists used genes from a scraggly Peruvian plant
growing in the wild to create a fatter, juicer, more disease-resistant
tomato. And in most cases improved varieties combine many
genes from different parts of the world.

My own hometown newspaper, the *Washington Post*, reports
such fascinating items in its Monday, Page 2 "Science Notebook."
I have done feature stories showing why a hybrid seed farmer in
Murdock, Nebraska, worries about the loss of plant species in developing countries. But hard news is difficult to come by.

The problem is not simply that slow trial-and-error genetic ad-
vances don't conform to conventions for breaking, front-page
news. The loss of species and its consequences aren't any easier to
chronicle.

The routine, irreversible loss of species is beyond dispute. Unique plant species sometimes exist on plots of land so small that pioneers can burn them into extinction in an afternoon of forest clearing. It is easy to *imagine* how much could be wiped out in tiny Costa Rica, which loses about 250 square miles of forest a year, but scientists don't *know* the losses with any precision.

To start with, the total number of the world's plants and animals is itself guesswork. Estimates run from 5 million to as much as 50 million. Scientists have described only 1.4 million. "Since most estimates of extinction are based on extrapolations, the lack of precise estimates of total numbers has led to considerable imprecision regarding extinction rates," according to *Conserving the World's Biological Diversity*, a book published in 1990 by several international organizations. "But the fact remains that basic knowledge of the organisms that make up most ecosystems, especially in the tropics, is woefully inadequate." Estimates of worldwide species loss range from 1 species an hour to 100 a day.

Mark Hertsgaard, writing in *Rolling Stone*, argues that the press should be as assertive as Walter Cronkite was when he closed his nightly news program with "and that's the way it is, the 23rd day of American hostages in captivity." But what would Dan Rather, Cronkite's successor at CBS, say about threats to biological diversity? "And that's the way it is, another day of losing somewhere between 24 and 100 species, one of which *might* have cured the common cold."

SCIENCE IS ONE part of the biodiversity story. Fidel Mendez is another.

For many years Don Fidel was a fisherman on Costa Rica's Pacific coast. When the fish began to disappear, he reached for the time-honored Costa Rican dream—a farm of his own. With 147 other families he invaded land near La Selva, and after lobbying the government to pay off the owner (another Costa Rican tradition), Don Fidel settled down to farming.

Walking through the ankle-deep mud along an ugly road cut through the forest, he talks with pride of protecting trees. They are an investment, he says, to be harvested as needed. He can name different varieties and is experimenting with 10 species that might have commercial applications.

Don Fidel also talks about his struggle to meet immediate needs. He grows cassava, ornamental plants and annatto, used in food coloring. All three, which Costa Rica exports, bring him cash. But world markets fluctuate, and tropical soils are not suited to farming. Farmers typically report smaller yields each year. In times of trouble, Don Fidel may have to cut and sell his trees just to eat. "By subsisting today, I know I destroy the future of the forest and the people," he acknowledges. "But I have to eat today."

Don Fidel's story is played out over and over by individuals and by nations. Difficult trade-offs between growth and the environment are a daily occurrence and are often complicated by hidden international factors. Journalists face an enormous challenge in illuminating these choices and their global implications.

An essential part of the problem lies with traditional views of progress. The exploitation of resources, rather than conservation, is the established predicate of development. Settlers of the New World sought to tame hostile environments; they introduced new plants and animals and destroyed much of the native flora and fauna. Charles Darwin, the father of modern biology, did not deviate from this norm. He regarded the loss of plant and animal species through natural selection as a healthy occurrence, not at all perilous.

No group of people more relentlessly altered their environment than North American settlers. Eighteenth-century European visitors commented on the unsightly aftermath of settlement in thick forests on the east coast of the United States. Ugly tree stumps surrounded farmhouses. Corn and wheat grew right up to front doors. Americans, wrote an Englishman, had an "unconquerable aversion to trees."

Economic systems, whether driven by Karl Marx or Adam Smith, have created record-keeping systems weighted toward exploitation. National accounts, such as used to calculate gross national product (GNP), record depreciation of plant and machinery but don't reflect depletion in a nation's supply of water, soil, air. In some cases environmental cleanup costs are counted as productive contributions to national output, when they should at most be seen as zeroing out past insults to the environment.

"Although the linkages between economics and environment are absolute, these linkages are not emphasized either by the economic or by the environmental professions," World Bank environmentalist Robert Goodland observed in *Race to Save the Tropics*. "The economics of natural resources remains a minor, unpopular theme in orthodox economics today."

Journalists can look to the U.S. Commerce Department each month to keep score on trade. Economic scorekeeping on the environment is in its infancy. Whenever fines are enforced, U.S. factories do a better job of monitoring pollution emissions. World Bank economists, among others, have begun to find ways of assessing a country's deforestation, oil depletion and soil erosion and of deducting that cost from GNP. But some aspects of the environment are not so easily measured. How, for instance, does one price a sunset? And, more to the point of this essay, how does one determine the worth of an individual species? The value of particular genes in a particular species can change. Horses with genes for strength were desired a century ago when farmers used them to plow fields and pull buggies, but not now. Moreover, one cannot consider species alone; they exist as part of an entire ecosystem.

"We don't know what we are losing when forests are cut down," said Dick Van Sloten of the International Board for Plant Genetic Resources at a 1991 meeting on biodiversity. "It's just that simple."

Without sound data, trade-offs between environmental protection and economic growth are difficult to calibrate. Everyone agrees that species loss is bad over the long haul, just as they agree that tropical deforestation damages the atmosphere over time. But how bad compared with providing more food right now for Fidel Mendez and his family?

Increases in the volume of world trade, the emergence of global stock trading and ever easier travel between countries further complicate assessments of environmental losses and gains. The environment can't be seen in isolation any more than a species can survive apart from its ecosystem, as Costa Rica's recent economic problems demonstrate.

In the late 1970s Costa Rica saw world prices for its commodities—like coffee—fall; prices for imports—like oil—rose. Large debts to foreign commercial banks added to its woes. Declining GNP jeopardized its national system of social services, the envy of Latin America. Some people also worried that economic troubles threatened the country's long record of political—and democratic—stability.

The traditional solution is economic growth, which in Costa Rica's case has meant increasing exports. The imperative to sell more abroad, however, subtly works against biological diversity. Costa Rican biologist Rodrigo Gámez explains why: There is "increased deforestation to raise timber exports (while forests remain); intensification of livestock ranching; and development of coffee, banana, cotton, rice, sorghum, oil palm and other types of higher commercial plantations." Corporate agricultural practices "flatten all the diversity you find there, not only in terms of species but even in topographic diversity. Bulldozers and planners just flatten out everything, and along with that goes not only topographic, microclimatic diversity that you require for a particular species, but the virtual disappearance of all other variability."

U.S. citizens face trade-offs too. Should they give high priority to guarding Costa Rica's genetic diversity, or should they emphasize its repayment of commercial loans to U.S. banks? The choice is never quite that clear-cut, of course, but that only makes the trade-offs more difficult to identify and manage.

The gold rush at Costa Rica's Corcovado National Park in the late 1970s is an example of the unpredictable loss of species that comes with proliferating global interactions. Located on the Osa Peninsula, Corcovado has one of Central America's best lowland rainforests. For years only a handful of poor miners bothered to search it for ore. Then foreign-owned banana companies pulled out of the neighboring area, causing widespread unemployment. Next, world gold prices soared. Waves of miners rolled into the park, especially its southern part. By the time the government evicted them, a government report noted, clear rivers and creeks, once full of animal life, were "liquid deserts."

Well-publicized debt-for-nature swaps have put land under protection in exchange for reducing foreign debts in Costa Rica and other countries. But stories about such deals are not complete without underscoring the continued vulnerability of these preserves. Unless economic opportunities exist outside the parks, people will invade them. To survive, both tropical forests and people must act in harmony. "If you don't have the right social and economic forces in place," says W. Franklin Harris, who helped manage a study of biodiversity research for the National Science Foundation, "you are doomed to failure."

Achieving harmony necessarily raises thorny social and political questions. One way to save germplasm is in gene banks, which have been set up around the world. But as important as these collection centers are, it is still essential to have plants and animals in their original settings, where they continue to evolve. In the case of plants and animals used by primitive tribes, the scientific ideal is for Indians to live as they always have. The problem is that Indians may have other ideas.

Likewise, developing countries often decry global environmental pressure to save the forests as neo-imperialism, an encroachment on their sovereignty, not to mention hypocritical in view of industrialized nations' raping of their own environment to acquire refrigerators in every house and two cars in every garage. As is well known, industrialized nations' relentless burning of fossil fuels contributes vastly more to global warming than does the destruction of tropical forests.

Third World resentment surfaces vividly with regard to biodiversity. Many in industrialized countries argue that free access to Third World genetic material is in the world's best interests: It facilitates scientific inquiry and the commercial use of genes. Developing countries, however, complain that they must bear the costs of environmental protection *and* pay industrialized countries' prices to get back their own genes in hybrid seeds or new medicines.

"Our forests are fulfilling a function for the world," Gámez says. "So it is for everybody's benefit to preserve these forests. But if we cannot cut these forests, or we should not cut these forests, we have to live on something."

The environment is a prime candidate to replace East-West confrontation as our number one national security consideration—only the issues will be more, not less, complicated as people give them more thought. The Persian Gulf crisis beginning in late 1990 was about natural resources. The obvious part was oil but, ultimately, genetic wealth was at stake as well. Many developing countries depend on imported oil and have higher energy import bills when supplies are curtailed. Acquiring money to pay for oil requires more pressure on fragile Third World economies and—you guessed it—on their environments.

"Policy-makers thus need to apprehend not only the range of accumulating problems but their multiple interrelationships as well," Norman Myers, a scientist, argued in *Foreign Policy*. When they do, "sustainable development," a tricky term for getting growth and conservation to go hand-in-hand, will be as important to strategists as missile "throw weights."

"I think the whole planet is caught in this problem of scarce land and growing numbers of people," one Costa Rican environmentalist told me. "If we want to make it to the 21st century, we need to find the other frontier—efficiency and sustainable development."

Such ideas could give rise to an age of "survival alliances" based on environmental considerations. But it won't be easy.

FOR EVERY COMPLEX problem there is always a simple solution, H. L. Mencken once observed, and it is always wrong. So it is in managing a world in which species loss in far distant countries shapes our lives. And so it is for journalists, who have to make sense of the world in which the environment has become a national-security issue.

Some reporters and editors have called for relentless coverage of the environment. That objective seems beyond dispute. But what about making an emotional commitment that calls for news reporters to be advocates?

As citizens of an environmentally besieged world, journalists should be introspective, as *New York Times* reporter Matthew L. Wald was in a 1990 Earth Day article. He observed that the story, running on two pages of newsprint, used "400 evergreen trees, mostly spruce, that covered about two acres." There is room for editorials and for the reporter who suggested in *Editor & Publisher* that journalists use the backs of press releases as scrap paper and, better yet, complain to PR people when they seem to be wasting paper.

But emotional declarations of faith can become part of the problem. "Fashionable alarmism may eventually create a Chicken Little backlash: As the years pass and nature doesn't end, people may stop listening when environmentalists issue warnings," journalist Gregg Easterbrook wrote in the *New Republic*. "The tough-minded case for environmental protection is ultimately more persuasive than the folk song and flowers approach."

Scientists are convincing when they say that we have global ecological troubles. But we don't know enough yet about the details. Put in other terms, environmental issues are too important for journalists not to live up to their pesky skepticism. They have to ask "so what" questions, even when doing so seems rank heresy. Even more than that, though, journalists have to ask "then what" questions. What are the consequences, the connections, between the environment and the rest of our lives. The two are woven together, but the templates that help reporters easily turn a local fire into a news story don't exist for the global environment. Journalists have a lot of learning and a lot of explaining to do.

We need specialized science reporters who understand biodiversity and the nature of greenhouse gases. But if connections with everyday life are to be made, everyone should cover the environment. "Educate reporters on all beats, be they national security, finance or local politics to recognize the environmental aspects of their stories," Mark Hertsgaard has suggested.

Many global environmental dilemmas remain as obscure as nighttime rustling in a tropical forest. In exploring them, journalists shouldn't look for neat paths. As one learns in the Costa Rican jungle, this is the time to complicate the story.

SCIENCE,
SOCIETY
and the
MEDIA

Stop and smell the flowers: Earth Day, 1970, New York
Courtesy Associated Press

SHARON M. FRIEDMAN

Two Decades of the Environmental Beat

When the *Seattle Times* won the 1990 Pulitzer Prize for national reporting for its coverage of the Exxon *Valdez* oil spill, it marked only the third time in Pulitzer history—and the first in 11 years—that an environmental story had been so honored. Not coincidentally, perhaps, one of two 1990 Pulitzers for public service went to the *Washington* (N.C.) *Daily News* for its coverage of a local water contamination scandal.

The media's renewed interest in the environment has made it *the* topic of this decade, heralded by *Time*'s decision to name the Earth its "Planet of the Year" for 1988. With that issue, *Time* made a commitment to increasing its coverage of environmental concerns, and other leading media organizations have made similar commitments. The Cable News Network has given environmental coverage high priority, and National Public Radio regularly provides strong coverage that weaves together the many disparate strands that make environmental reporting so difficult. As well, environmental stories receive frequent front-page treatment from the likes of the *New York Times* and the *Los Angeles Times*. The *Los Angeles Times*, in particular, is often cited as the American newspaper with the best environmental reporting and has moved to make logical connections with other coverage areas, such as business, by appointing an environmental specialist to that beat.

Publications that once gave little or no attention to environmental reporting, such as *BusinessWeek*, now have staff devoted specifically to that purpose. Elsewhere, according to a recent study conducted by the Scientists' Institute for Public Information, 138 small newspapers give more coverage to environmental issues than they had done, even 1 or 2 years earlier. In media organizations large and small, environmental beats are being established once more after having vanished for a number of years.

Among other recent developments are these:

• In May 1990, the *Rocky Mountain News* began a daily page devoted exclusively to environmental and science stories. It is believed to be the first of its kind in the nation.

• In July 1990, KOB-TV in Albuquerque launched an "E team" after market research showed the environment is the issue of greatest interest to its viewers. The team is seeking to produce one environmental story a day.

• In fall 1990, the Public Broadcasting System aired "Race to Save the Planet," a 10-part series on environmental issues that was filmed in 30 nations on all 7 continents.

These are just a few examples of the ambitious environmental reporting now going on at newspapers, magazines, television stations and other media here and abroad. And the trend does not appear to be abating. "We feel that environmental issues are going to be the issues of the '90s," said Spencer Kinard, vice president and news director of KSL-TV in Salt Lake City. Gina Stuki, a senior account executive at Audience Research and Development, a market research firm in Dallas whose clients include 100 television stations across the country, added, "In half of the markets we cover, the environment is of more interest to viewers than health and medicine. . . . We saw the trend in coastal regions first, but now it's spreading across the country."

Three new national magazines devoted to the environment—*E* magazine, *Buzzworm* and *Garbage: A Practical Journal for the Environment*—have begun publication during the past 2 years. There's even a new national environmental magazine for children being published, *P3, The Earth-Based Magazine for Kids*. The publishing industry, ever in search of ways to sell more books, has been churning out volumes on a wide range of environmental topics. John Javna's book, *50 Simple Things You Can Do to Save the Earth*, remained on the best-seller list for several months following its publication. At least 12 magazines published special Earth Day 1990 issues, including *Harper's, New York, Newsweek, Smithsonian, Sports Illustrated, Time* and even *TV Guide*. The *Media Monitor*, published by the Center for Media and Public Affairs, looked at network evening newscasts and weekly newsmagazines and found 595 news items on the environment during 1989. Of these, 453 stories were broadcast on ABC, CBS or NBC, and 142 articles were found in *Time, Newsweek* and *U.S. News and World Report*.

There is little doubt that the number of reporters assigned to the environmental beat and the amount of airtime and pages broadcast stations and newspapers devote to this subject have increased significantly during the period 1988 to 1990. Yet despite these efforts, the environmental beat of the 1990s is not very different from what it was in the 1970s. While quantity may be up and environmental topics different and more varied, the quality of environmental coverage presents many of the same problems it did 20 years ago. There are other similarities as well. No one knows now—just as no one knew then—how many environmental reporters there are working in the mass media or just what topics fall under the rubric of environmental reporting. Where does one draw the line between science and environmental reporting, or between political and environmental reporting?

In a 1972–73 study by William Witt of the University of Wisconsin, environmental reporters said they felt that their beat was inadequately covered and blamed other reporters and editors for ignorance and indifference. They also blamed newspapers for not supplying enough staff, time and space. These reporters castigated the press for too much crisis reporting, too much sensationalism, and too little interest in following the complexities and development of most environmental issues. They felt that reporting dealt too much with environmental problems and too little with solutions and, finally, that environmental reporters needed more education both in science and government, because the beat was a combination of the two.

During the 1970s other writers criticized environmental journalism for crisis reporting, with one calling it "eco-journalism"— "the practice of reporting ecological crises by ignoring, treating as unimportant, or mishandling the evidence on which the crises are based." Local newspapers were accused of practicing "Afghanistanism"—concentrating on geographically distant environmental problems rather than those in their own backyards—a practice that contributed to the belief that local environmental problems were relatively unimportant.

In the 1980s other problems began to crop up, and in 1983 I discussed some of them in an article for *Environment* magazine. At that time the environmental movement was in its doldrums. Many environmental beats had disappeared or had been institutionalized—absorbed into general news, energy or science reporting— and the few environmental reporters who remained often were starved for space, time and front pages. Unlike many other topics, the environment was plagued both by enormous complexity, tied as it was to a range of economic and political issues, and a great deal of scientific uncertainty. As if these problems weren't difficult enough, they were made worse by journalism's need for a hard news peg and short deadlines, by the lack of adequate sources and the tendency toward crisis reporting.

There was no question in the early 1980s that environmental

issues were complex and getting more so. We began to realize that environmental subjects were multifaceted, involving not just technical information but also legal, financial, political and social considerations. Much of what needed to be covered affected human health and involved evaluating costs and benefits and government regulation. Moreover, environmental concerns usually involved a long and tenuous string of interrelated concerns that all had an impact on people's lives. To cover a story properly, the environmental reporter needed to deal with all of them. As well, there were almost always uncertainties about scientific data. Experts disagreed about interpretations and applications of research, or how to manage environmental trade-offs. Neither complexity nor uncertainty could be covered effectively using traditional reporting techniques.

These problems are not very different from those Witt identified in 1972–73, and they are still with us today. Studies of various environmental issues done over the last 10 years substantiate and reinforce these coverage criticisms. Narrow choice of sources, avoidance of scientific and technical factors, and lack of in-depth reporting are key problem areas. So is crisis reporting of environmental issues, which still dominates and reflects the media's preference for conflict and drama.

Despite criticism, journalists have not done much to establish new and better sources: Government officials are far and away the major information source used in most environmental stories. In 1983 I found that reporters, particularly those on general assignment, either did not look for objective and knowledgeable sources or had trouble finding them, and depended too much on local officials. *Media Monitor* points out that in 1989, for environmental coverage on television evening newscasts and in weekly newsmagazines, government officials accounted for 32 percent of the sources used—twice the percentage of the next most frequently used group of sources. Environmental reporters attending an American Press Institute (API) workshop in June 1989 said that government officials were their predominant source, followed by environmental groups. Industry officials, scientists and private citizens ranked far behind.

There are a number of reasons for this dependence: Government officials are considered credible and authoritative; they often are easy to reach, and many are used to talking with reporters. Journalists rightly point out that they need to talk to government sources, as they are the ones either administering the laws or regulating environmental hazards. However, using government officials as sources too often leaves a reporter open for manipulation. In Washington, in particular, the practice of reporting environmental news by taking government agency handouts is a serious problem. Jim Sibbison, a former press officer for the Environmental Protection Agency (EPA), charged in 1988 that a principal occupational hazard in environmental reporting from Washington is "relaying to readers self-serving statements by EPA officials as truth."

Lack of choice in sources also is a major problem, particularly for general-assignment reporters who cover environmental issues on a part-time basis. A number of studies have found that they do not know where to go for information and objective evaluations. Often they do not have the time to cultivate sources, even local ones, who can help them interpret technical information. In one brief survey I did, only 1 out of 60 Pennsylvania reporters was aware of the Media Resource Service, run by the Scientists' Institute for Public Information, which has a computerized list of more than 20,000 technical experts willing to answer reporters' questions on scientific and technical subjects. Others had little idea of the multitude of national environmental and scientific organizations they could call on for assistance.

At the heart of most environmental issues lie scientific or technical concerns. Just what was the risk to children who ate apples impregnated with Alar? How much exposure to radon is dangerous? Does coal tar leaking into a sewer put a community at risk? Is global warming happening or not? All of these questions have answers steeped in technical data and interpretation.

Typically, environmental reporting avoids the technical aspects of such issues. For example, in a study that Carole Gorney, Brenda Egolf and I did of radiation reporting in U.S. newspapers and on the major television networks during the Chernobyl accident, we found important topics with scientific overtones frequently missing from the coverage. These included different types of radiation or levels that could be harmful to people, food and water supplies; how specific radioactive elements affect humans and the environment; differences in effect between short-lived and long-lived radiation elements; and long-term effects of radiation, such as cancer, on people. Studies on Bhopal and Chernobyl, among many others, have called for more in-depth coverage to provide perspective, particularly for stories that deal with environmental risk issues. Bhopal was portrayed as the kind of event that people most dread, with little context and with certain essential facts omitted. Lee Wilkins of the University of Missouri notes that U.S. reporters did not put the Bhopal event in the framework of larger issues that surrounded the plant—the need for jobs in the developing world, the need for food in those same countries, and the role that chemical pesticides and fertilizers play in fulfilling those needs. Coverage of Chernobyl, besides lacking in-depth examination of major radiation factors, also lacked background on why the Soviet Union was building so many nuclear reactors, why this particular design had been chosen, how power needs affected the Soviet economy, and the politics of nuclear power issues in Europe.

Providing such background information empowers readers and viewers, giving them information with which to make decisions. Control over environmental risks and hazards is a major factor for citizens, who are more apt to accept a risk if they feel they have some degree of control over it. But people seeing only facts without context in hazardous situations may decide that they are helpless to intervene or change a situation, and therefore may not participate in the debate. Traditional news gathering, with its emphasis on heroes, major players and "big" events, encourages such attitudes. What was true for Bhopal and Chernobyl is also true for reporting on radon, toxic chemicals and many other environmental and health hazards. The event takes precedence, and perspective and follow-up are lost in the process of only reporting the news.

WHILE IT IS hard to tell which of the coverage problems we have been discussing is the most serious, the related problems of crisis reporting and outright hype certainly get the most complaints.

Oil spill coverage for 1989 is a good example of what some would call hyped, or overplayed, environmental news. The *Media Monitor* reported that of the 595 news items that appeared on the television network evening newscasts and in weekly newsmagazines, 238 were about oil spills, with 189 of these about the Exxon *Valdez* spill. (The next largest category was animal protection, with 52 stories.) The Exxon *Valdez* dominated nearly half of television's total environmental coverage for the year, while important, long-range problems like agricultural runoff, overfishing, pesticide use and the solid waste explosion were neglected.

Coverage of the 1989 Alar furor demonstrates the problems that media hype can cause. According to the *Wall Street Journal*, stories on the charges made about Alar on apples by the Natural Resources Defense Council (NRDC) and the subsequent formation of its Mothers and Others for Pesticide Limits found their way into hundreds of media outlets as varied as "60 Minutes," "The Phil Donahue Show," the *Washington Post* and *People*. This massive coverage resulted in the banning of U.S. apples by Taiwan; elimination of apples from school lunch programs in cities like New York, Chicago and Los Angeles; and a significant decline in apple sales across the nation. Yet while the coverage was extensive, it was hardly adequate. Most of it focused on the NRDC charges (made easier by the NRDC's enlistment of Meryl Streep as its spokesperson on this issue), EPA explanations, industry rebuttals and economic aspects, and only a fraction—about 15 percent of the stories—included some type of risk figure to help readers put the situation into perspective. Numerous officials and even environmentalists have charged that the Alar issue was blown far out of proportion.

That environmental coverage has not improved much since the 1970s is particularly regrettable because environmental issues have changed. Twenty years ago we were not really considering how to communicate risk to people—how to explain their chances of getting cancer when exposed to 6 parts per billion of dioxin coming from a trash-to-steam incinerator, or mild electromagnetic fields that envelop their homes, or radon in their basements. Twenty years ago we were not dealing with what some scientists were then telling us about the greenhouse effect and other global environmental hazards. Now we have to worry about deforestation and desertification in places like Brazil and Thailand, where Third World poverty is a principal factor in these processes. Environmental issues are now being debated at the presidential and prime ministerial levels, yet the media are covering them much as they did in 1970.

Of course there has been some progress. Some journalists, mostly from the larger publications and cable and broadcast networks, are doing in-depth work on global, national, regional and local environmental issues; and the scope of their coverage—from Third World deforestation to First World industrial development—far exceeds anything they have done before. Investigative reporting has exposed poor environmental practices and ineffective laws. According to Jim Detjen, a reporter for the *Philadelphia Inquirer* and president of the Society of Environmental Journalists, some U.S. reporters have begun to travel to the rainforests and other foreign locales to view environmental problems and report on them firsthand. Others are starting to work with their colleagues in Asia, Eastern Europe and Brazil to share information, techniques and sources. And environmental journalism is on the rise in other countries too. With the help of the United Nations, Asian journalists have formed an 11-nation

Asian Forum of Environmental Journalists to encourage and co-ordinate efforts for better environmental coverage in their countries.

But will this be enough? Reporters seeking more time and space to provide in-depth, comprehensive coverage are still faced with editors who ask why this beat is different from any other. At the same time, the range and scope of environmental issues have raised new and thorny questions about what issues to cover and the rise of advocacy journalism. "Environmental reporters must help the public decide which environmental problems deserve the most immediate attention," says Morris "Bud" Ward, editor of *Environment Writer*, a newsletter published by the Environmental Health Center of the National Safety Council. But in doing this, where does objective reporting end and advocacy reporting begin?

One reporter attending the API workshop in June 1990 noted that the line between objective and advocacy reporting keeps getting blurred. Other reporters have forthrightly suggested that environmental reporters disregard "objectivity," that traditional journalistic methods are both impotent and imprecise for environmental coverage, and that the nation and the world can no longer afford them. Mark Hertsgaard, a contributing editor for *Rolling Stone*, said recently, "The media are in an absolutely pivotal position in the race to save the planet. The next challenge is to move now from saying that the environment is a difficult problem to the next stage, which is to diagnose the specifics of the problem and the means of action."

One such means, certainly, involves the use of increasingly sophisticated information-gathering technologies to help journalists sort through the multitudes of reports, facts and figures on environmental issues. A reporter might, for example, use an electronic data file to match records of hazardous chemical emissions from industries on the EPA's Toxic Release Inventory with records of corporate donations to politicians' election campaigns. Such projects are expensive, however, and current trends in the newspaper industry—trimming back on newsholes in a quest for greater profit—endanger not only special projects and in-depth reporting but also the technologies that make them more effective.

Finally, risk reporting remains an occupational thorn for environmental journalism. Cristine Russell, special health correspondent for the *Washington Post*, has accused the media of doing a poor job of putting health risks into perspective. While cancer gets a great deal of attention, she says, reporters need to ask more questions about more subtle risks such as those to the central nervous and reproductive systems. "Increasingly," she says, "journalists are the brokers for information. We will have more information to sort through and will have a more difficult time helping readers or viewers understand what it means."

MORE AND MORE people, including those in the profession, are calling on environmental journalists to change, to become educators rather than just providers of information. This means different things to different reporters. To educate, in my mind, means that environmental reporters—with support from their editors—provide depth and context, and frame information in such a way as to lead readers or viewers to refined judgments. Educational reporting helps analyze what effects the basic facts have on citizens and their communities, or perhaps on the world. In some instances it can even show people how to bring about change.

I personally do not think environmental reporters have much choice but to educate if they are going to keep pace with the public's growing level of sophistication about environmental issues. Environment is becoming such a predominant issue that it will eventually permeate almost every beat. Every reporter, not just specialists, will occasionally be writing about the environment from some perspective. But while any good reporter can provide the facts, it will be the environmental reporter's job to provide the context and background that readers and viewers need to understand the issues. If this does not happen—if the majority of environmental reporters continue their present practices—there is a good chance that the beat will once again become absorbed or "institutionalized" as part of what other reporters cover under politics, business or health, as it did in the late 1970s and early '80s. It is up to environmental journalists to make sure that the beat goes on.

JOHN BURNHAM

Of Science and Superstition: The Media and Biopolitics

On Halloween weekend in 1948, 20 people choked to death on the smog in Donora, Pennsylvania, an industrial town in the mountains near Pittsburgh. Those who died and others who had to go to the hospital were already ill with respiratory diseases, and most were elderly. The authors of the initial press reports characterized the incident as they would have an industrial accident, with a minimum of human interest material and with much quoting of officials, who of course were uncertain as to the cause and called for an inquiry. Even the newsmagazines did not get much further with an essentially minor story, although they did report one harbinger of the future: A Donora health official who blamed local industrial air pollution for the deaths asserted, "It's plain murder."

Americans who had long wanted to clean up the air around Pittsburgh, Los Angeles and other cities were soon using Donora as a symbol of the danger that poisons in the air posed for all citizens. In 1962, biologist Rachel Carson published her classic *Silent Spring* and greatly extended the knowledge that many people had

The U.S. Post Office issued its Rachel Carson commemorative stamp in May 1981, 19 years after the publication of Silent Spring
U.S. Postal Service

about the extent to which human activities were endangering the life around them, in the water and soil as well as the air. Donora was soon well established not in the category of "accident" but as the incident that precipitated the environmental "movement." During the 1960s Congress had enacted clean air and water acts and a wilderness act. By 1970 the great media event, Earth Day, was inaugurated to further raise the alarm—and consciousness— about the ways in which human beings were poisoning their planet, and it was counted the largest political demonstration to that time.

Altogether, the category for environmentalism had become not aesthetics or science but politics and public affairs. In 1990, when millions of people celebrated the 20th anniversary of Earth Day, a majority of Americans, according to a *Newsweek* poll, thought that Earth Day was mostly media hype.

In the course of a few decades, then, environmentalism changed. It started as learning about human interactions with nature, and it became public affairs and media hype. In the transition, a special, well-recognized institution was involved, namely what was known as the popularization of science, for people believed that environmentalism depended upon public understanding of science. The movement was in fact based on ideas of ecological balances taken from biology as well as on specifics from the broad medical specialty of public health. But ideas about what constituted "public understanding of science" and how to cultivate it had already begun changing during the 20th century. Those changes in the popularization of science help make the course of environmentalism understandable because they involved a similar transformation from science to public policy/politics and media hype.

BEFORE 1800 ANY educated person could read and even contribute to science. Beginning in the 1830s, however, the subject matter of natural philosophy (the physical sciences) and natural history (the biological and earth sciences) fell into the hands of professional and specialist researchers who produced such technical knowledge that someone had to simplify, summarize and interpret for the general intelligent public the teachings and findings of investigators.

At first the summarizers and explainers were, to a substantial extent, scientists themselves. Not all scientists chose to popularize, but a remarkable set of what became known as "the men of science" spoke to the public and wrote for the media of that day about mathematics, astronomy, physics, chemistry, botany, zoology, physiology and even psychology.

These "men of science" (including a number of women) had an agenda. They wanted on the one hand to teach people to appreciate nature and satisfy their curiosity about natural objects and processes; on the other hand, the popularizers were working to demystify the world and introduce a healthy skepticism. The paradigm of demystification was Benjamin Franklin's demonstration that lightning was merely static electricity, a demonstration that brought a terrifying phenomenon into a world of rational understanding. The enemy of popularized science was superstition, which was authoritarian and irrational. As one of the chief men of science, David Starr Jordan, the ichthyologist president of Stanford, wrote early in the 20th century: "The two great functions of science are broadening of the human mind, its release from tyrannies of ignorance and of self-constituted authority. Hence arises the second great function, the use of all knowledge needed for human health and efficiency and for all phases of the great art of the conduct of life."

Such popularizers as Jordan therefore tied science to progress and traced how human civilization flourished when people acted rationally and society encouraged curiosity. Eventually popularizers hoped to teach everyone to follow the restraint and calculation of the scientific method—"the great art of the conduct of life." Such uplifted people, the men of science assumed, would, like themselves, seek high culture and have puritan consciences. The religion of science that the men of science preached in the late 19th and early 20th centuries therefore involved a world view: All knowledge was interesting because it fitted into a rational and empirical view of the universe. An animal or plant could be seen adapting; a new star extended what we knew of galaxies, space and gravity; a chemical was related to others in the periodic table or interacted with other chemicals in a way that had meaning in patterns of reactivity.

One other attribute of the men of science was of special significance: They sold large parts of the public (and themselves) on the idea that the truth of the scientist was superior to other truths because the scientist was self-effacing; he or she was objective. Using the scientific method, they had succeeded, as economist Wesley C. Mitchell put it, "in emancipating themselves from the misconceptions and prejudices prevailing in their social groups."

Yet as the decades passed in the 20th century, the general public heard less and less from scientists. Instead, almost all scientists left to journalists the job of popularizing science for the general public. Why, after all, in a specialized world, should scientists intrude into the specialized domain of informing the public, especially after journalists, as sociologist Michael Schudson has pointed out, had developed a new respect for facts?

Indeed, in the United States a specialized corps of science writers materialized after World War I to furnish the public with authoritative accounts of science. The science writers often worked closely with scientists and by midcentury tended to reflect the funding and publicity efforts of leading members of the research establishment.

In addition to being mostly uncritical of science and scientists, the midcentury science writers were burdened by three conventions of journalism. First, like other journalists, they emphasized facts—to the point that the science they portrayed was a parody: One discovery followed another, but not in any context, just something new, i.e., news. Second, in order to affirm the practical need for science, the writers followed the lead of the scientists and justified research because it had practical results. The consequence was that everyone confused science with technology. In this type of publicizing, the reason that science was good was that it brought refrigeration and plastics, not the scientific method. Finally, health concerns dominated science news. New cures such as "the miracle drug" penicillin dominated the media as holdovers from the sensationalism of the old yellow journalism. The science writers, who were neither stupid nor unperceptive, joked about the "new hope" school of journalism in which each discovery brought new hope for sufferers from one disease or another.

It was into this world of journalistic popularizing that the environmental movement came. The movement came as a disrupting and disturbing force, for environmentalism did not always fit the model of science popularizing and instead harked back to the sensationalism and opinion that marked an earlier and discredited journalism.

The trouble with environmentalism started because it was based in biology. Biology had always been problematic for popularizers, even aside from the sensitive issue of evolution. In 1928, for example, biologist Oscar Riddle of the Carnegie Institution of Washington noted with envy "those who have taken the atom apart . . . and others who have presented us with a quite new and gripping conception of the extent and structure of the universe"; in contrast, life scientists like himself had to try to encourage public confidence with phenomena that lacked exactitude and rigid predictability. (Riddle was at the time trying to explain the place of internal secretions in evolution.) Biology did not involve just complexity but also a great deal of inexactitude. Even in quantitative areas such as genetics, predictions were only statistical approximations, not, as in physics, mechanical certainties the exceptions to which (like the uncertainty principle) were merely theoretical. In 1979, Phillip J. Tichenor and other journalism educators referred to environmental reporting as the "journalism of uncertainty."

As knowledge about the endless interdependence of living systems grew, popularizers of biology therefore concentrated ever more on meaningless facts—discoveries—out of context. The distribution of flora in an exotic environment or the metabolism of a mold could not compete in the media with a new autogiro or even a radioactive element.

An additional problem for journalists was that environmentalists tended to criticize the American science establishment—indeed all of the establishments. Writers who were at home publicizing cures for cancer discovered in outer space, as one journalist satirized his own specialty, were decidedly unprepared to deal with the underlying assumptions of ecofreaks whose hostility targeted not just chemists and the chemical industry but, implicitly, capitalism in general. Environmental enthusiasts calculated, as historian Samuel P. Hays points out, not the costs of production but the place of any activity in the economy of nature.

Environmentalists specifically blamed technology for endangering the environment. It was one scientist's opinion (quoted by Dorothy Nelkin in her 1980 book *Selling Science*) that "the press often seems intent on showing how modern technology, including the chemical industry and nuclear power plants, is poisoning America." Yet the usual popularizations of science had portrayed technology as the end product of scientific discovery, and in fact industrial scientists had customarily been among those most helpful in furnishing material to compliant journalists. Conventional science writers, accustomed to praising applied science, were therefore slow to jump onto the environmental bandwagon. Long after the first Earth Day, poll data showed that they and their editors still considered established business sources more reliable than citizens' groups.

Perhaps the most unhappy aspect of attempts to popularize environmentalism as science grew out of the fact that many of the supporters of the movement were fundamentally anti-scientific, or at least anti-rational. The imagery of endangered species, for example, was based on scientific facticity but with a focus on nature, not on science. The environmentalists' commitment to a romantic viewpoint, which hallowed nature and abjured attempts to manipulate and subordinate the environment, contravened the idea of control that underlay understanding and predicting natural phenomena. ("The chemical weed killers are a bright new toy," wrote Rachel Carson in a biting passage in *Silent Spring*. "They work in a spectacular way; they give a giddy sense of power over nature to those who wield them. . . .") Indeed, the environmentalists often held the same ambivalent attitudes toward the study of nature that romantics did in the early 19th century. Moreover, and confusingly, many of the passionate advocates of sometimes antiscientific environmentalism were well-qualified scientists.

From the popularizers' point of view, then, environmentalism was science—and was not science. The scientific content indeed came from biology, and the urgency was derived in part from public health concerns, long the staple of popularizing. But the news stories were not coming from the laboratories. "The majority of the reporters plying the environmental beat on today's dailies," reported journalism educator Clay Schoenfeld in 1979, "have come from the ranks of public affairs reporters, indicative of the fact that the environmental story so frequently surfaces in courthouses, capitols and judicial chambers." The truth was that environmental science involved not curiosity or even uplifting demystification, but social action.

While what passed for science writing in the environmental movement had roots in nature writing and outdoor writing—the aesthetic element of the older preservation and conservation movements—the urgency grew out of a late 20th-century concern: the self-centered self.

The old religion of popular science had involved altruism as well as self-denying objectivity. But 20th-century science writers took from the journalism of their day the idea that, in order to get the attention of the public, they had to make readers feel that they were personally involved, not as learners, but in a material way. Members of the public interviewed at the New York World's Fair at the end of the 1930s affirmed that they responded best to popularization of developments that affected them personally—not scientific theory, but inventions that would improve the performance of their cars. Within a generation, such attitudes greatly intensified in the age of narcissism and finally the "me decade" of the 1970s, leaving a mark not only on science popularization but on the environmental movement. There was, noted a science writer of that era, "an emphasis away from cold, indifferent science to that which produces some gain in the quality of life," and the implication was that the gain was immediate and personal.

As the environmental movement took shape in the 1960s, several basic concerns emerged: poisoning of the air, water and earth; exhaustion of the soil; depletion of other natural resources; and overpopulation/mass starvation. As social scientist Edith Efron has observed, by the beginning of the 1970s public health concerns had come to dominate the movement, and what she calls the new apocalyptics were saying that poisons in the air, water and earth, such as radioactive fallout, were affecting not only Americans but everyone in the world. Overpopulation and depletion of natural resources were possibilities for the future, but cancer was already present. It was a fortuitous coincidence that installments of *Silent Spring* first appeared in *The New Yorker* just after the thalidomide tragedy broke in the media, and for decades thereafter tales of man-made poisons in the environment, like those at Donora, made DuPont's advertising slogan of "better things for better living . . . through chemistry" sound like satire.

In the age of narcissism, then, environmental writers played up how environmental damage would affect each American personally and, in turn, what individual initiatives were necessary to forestall such damage. This tactic of stimulating personal fear continued in succeeding years to fuel the movement and to mark news reporting about it, as journalists struggled through the Three Mile Island accident and the dangerous-medical-wastes-on-the-beaches scandal ("Don't go near the water," warned *Newsweek* in 1988). Each new fact or discovery resonated with personal relevance rather than expanding understanding of the universe or encouraging a scientific approach to solving problems.

It was an ironically negative version of the more conventional science writing summarized in the title of a 1978 book, *Science Fact: Astounding and Exciting Developments That Will Transform Your Life*.

In the 1980s, while concern about poisons continued at a high level, the agendas for the future began to loom larger in environmental writing. Many factors were involved: reaction against the anti-environmental policies of the Reagan administration, a new sense of globalism, and immediate local problems of waste disposal, to mention some of the obvious.

But another factor was operating also: As journalists took over the task of popularizing science, they moved it into their own world. They were not neutral translators. Rather, they imposed on their public an agenda that competed with that of the men and women of science who wanted to instruct and demystify the world. The competition between the world of the media and the world of science and nature was mostly implicit, but occasionally signs of the rivalry broke through. As early as the mid-1950s a journalist blurted out in a roundtable discussion, "I think the doctors have no kick coming about us reporters, because 98 percent of the time I am writing about the wonderful discoveries and treatments they are making, and about 2 percent of the time their foolish excursions into pontificating upon what is good for the public and things like that." Clearly, the journalist did not want to be a party to making science a source of reform or uplift. Scientists were good only for furnishing disconnected facts. Journalists found others to pontificate—or did it themselves.

It was in the journalists' world that facts were disconnected in science writing. Within that world, reality consisted of what was in the media, which was and is chiefly the world of politics and personality. Many writers have described how the media reified themselves and how this process intensified in the age of television. (Science facts on television usually appeared in the form of natural history programs, but they were declining in numbers by the 1980s except for cable reruns. Insofar as scientists appeared in the entertainment sectors, communications scholar George Gerbner has reported that the more a person—regardless of demographics—watches television, the more hostile to science and scientists his or her attitudes. At the same time, in the spring of 1990, *TV Guide* reported that "among Hollywood's leading TV producers, ecology is the hottest thing going.") But in the case of science, media writers were implicitly disintegrating and so disestablishing and discrediting the alternative world of natural science. Earth Day 1990 commemorated a great media event, which had set off serious media coverage of environmentalism. It was the day on which the media discovered the ecological crisis, which for 20 years has served when other newsworthy crises were in short supply. Neither Darwin nor Carson was commemorated as such; a pseudoevent was.

Environmentalism, it turns out, works into the media world particularly well. Once the basic idea developed that every event is related to every other event by ecological links in "the chain of life," all other science became "facts." If the dinosaurs were wiped out 60 million years ago by an asteroid collision, that is a fact, or perhaps two facts, with implications for what can happen to large animals on the earth when air pollution reaches great intensity. But in the media the extinction of the dinosaurs does not involve methodology or even the debate over, say, cladistics and the mechanisms of evolution—much less skeptical and reductionistic explanations of life. (And as polls reveal, it has left most Americans still believing—as portrayed in the entertainment media—that human beings coexisted with the dinosaurs.)

In addition, by making science a part of policy-making, popularizers of science working with environmentalism have trivialized science by rendering it significant only in terms of what policy-makers do. Science educators are told to make people scientifically literate so that they can vote and speak intelligently upon environmental issues, not understand how the universe works. From this point of view, science, and particularly ecological thinking, is intended to have conventional policy implications, not extend rational and systematic scientific thinking to undermine myths of the late 20th century.

Finally, journalistic treatments of environmental matters have gutted science of content by reducing it not to a method for finding truth but to a conflict of authorities and even personalities quoted in the media. Aside from self-publicizing investigators seeking attention and funding, even well-intentioned "visible" scientists have been caught up with the journalists, as Dorothy Nelkin points out, in emphasizing the imagery, not the substance, of science. Emphasizing conflicting authorities assists substantially in shifting the environmental debate to arguing about what might happen in the future; the fate of the physical planet and the population becomes just a matter of opinion that will be decided on the basis of media presentation, not appeals to research and scientific method. The greenhouse effect, for example, has lent itself particularly well to trivialization by juxtaposing competing authorities, a process engineered by supposedly neutral journalists who are looking for conflict, not consensus. According to news reports and magazine articles, environmentalists speak of stressing the environment to the "threshold" past which nature cannot sustain life, of the unknown dangers of genetic engineering, and so on. From these warnings, however, the public learns not science, but that there are competing figures speaking on one side or the other, each trying to establish the most convincing authority and media presence. Indeed, astrologers can get equal time with the best investigators. The goal of reporters is to find and dramatize "controversy."

By bringing environmentalism into the world of media reality and reducing science to authoritative assertions, detached from method and the validating processes of the scientific community, modern journalists have popularized environmentalism into impotence except as a political—as opposed to a scientific or even educational—movement. (Nor do most of the timid and narrow specialists who pass for scientists in the late 20th century try to reclaim popular science or preach a religion of science, with or without the environmental movement.)

What started out as an unimportant story of an accident at Donora has, in journalists' hands, been safely derailed from the disturbing world view that the men and women of science once preached to the public, in which events of geology and biology, to say nothing of chemistry and physics, are real, and in which the scientific method determines validity concerning the world about us and even offers a set of values. Instead of seeking the authority of the scientific community—or even fairly primitive empiricism, as in drinking water fluoridation—many journalists, using journalistic standards, have made the judgment call that environmental science is all just a matter of opinion—or perhaps just "eco-chic."

The best of journalists are deeply concerned with the practices and institutions within which they have to work. They identify the problems with great intelligence, but their search for amelioration shows that basic improvement lies outside the world of journalism. They have at least trembled on the edge of recognizing that science is a moral as well as a technical enterprise. The moral fervor of the ecofreaks, as well as other changes beginning to unfold in the 1990s, suggests that with or without politics both journalists and scientists should find moral advocacy more acceptable than it was 20 years ago. Only if they act on that opportunity can science be reclaimed from a world in which disembodied facts and controversy are ends in themselves, rather than the means to come to terms with the natural world.

The world's first view of the Earth from the vicinity of the moon, taken by
U.S. Lunar Orbiter I on August 23, 1966
Courtesy NASA

ROBERT GOTTLIEB

An Odd Assortment of Allies: American Environmentalism in the 1990s

What is environmentalism, and where might it be going now, at a time when its popularity and influence have never been greater? As we enter the 1990s, with environmental issues racing to the top of policy agendas East and West, North and South, it is becoming increasingly clear that environmentalism stands at a crucial transition point, poised between different kinds of movements and differing interpretations of its past, present and future. The more dynamic yet less visible part of the movement, consisting of hundreds of local grass roots groups that are populist in spirit and heritage, has been a driving force for institutional change in key areas of daily life.

These groups contrast with the big, national, staff-based, expertise-oriented organizations such as the Sierra Club, Environmental Defense Fund and Conservation Foundation. These are the groups who get to speak in the name of "environmentalism" and who have been the primary focus of the media since the Earth Day 1970 events helped usher in a new agenda of environmental regulation and cleanup. Today, that agenda has become increasingly problematic in light of the continuing, protracted environmental problems we confront and the kinds of questions being raised by the local groups, questions that have forced a reevaluation of both the larger agenda and the interpretation of environmentalism associated with it.

Reflecting on Earth Day 1970, in fact, can be a useful starting point for such a reevaluation. This colorful and varied set of demonstrations, teach-ins, rallies, guerrilla theater and confrontations made reference more to the tactics and concerns of the 1960s and, beyond that, to earlier expressions of discontent about urban and industrial forces, than did the more narrowly defined agendas of such groups as the Wilderness Society or the Sierra Club. The concerns expressed at Earth Day 1970 were primarily about the quality of daily life—industrial pollution, the role of the automobile, consumption and waste, and the deterioration of the urban environment. Similar concerns had been articulated in documents like the 1962 *Port Huron Statement* of the Students for a Democratic Society, a seminal expression of that era. Moreover, the wide range of local groups that emerged during the late 1960s—from Get Oil Out in Santa Barbara (in the wake of the 1969 oil spill) and Pittsburgh's GASP (a community-workplace coalition built around the air emissions of the steel mills) to the New York City–based Urban Underground (consisting of disaffected urban planners) as well as Earth Day's convening organization, Environmental Action—came to define their agendas around distinctive urban and industrial issues and their enormous impact on natural and social environments.

These concerns also resonated historically in such pre–World War I forms as the female-led municipal housekeeping movement, which focused on issues like sanitation, public health, and food and nutrition; the muckraker-inspired urban reform organizations and settlement houses, which led the push for better housing, for new regulatory agencies such as the Food and Drug Administration, and for improving the harsh conditions of industrial labor; and the "sewer" socialists, who helped revolutionize municipal governance over the urban environment. These urban and industrial movements, though dealing with the questions of environment in the context of daily life, have nevertheless not been seen as part of an "environmental movement" in the same way that their contemporaries—the protection-oriented preservationists, led by John Muir, and the management-oriented conservationists, led by Gifford Pinchot—have been recognized as the forebears of contemporary environmentalism. The preservationists and conservationists, clashing over whether to set apart or more efficiently manage nature and its resources for protection against overarching urban and industrial influences, established the dichotomy between nature and society that came to be associated with the definitions of environmentalism. But it was precisely those daily-life environmental issues that were to reemerge more than a half century later as the dominant themes of Earth Day 1970. The broad sweep of the Earth Day actions, in fact, appeared to place such matters as air, water, food, garbage, energy, housing, transportation and toxics as priority items for an environmental agenda even as the descendants of Muir and Pinchot were articulating their concerns over a diminishing wilderness or an expanding population.

The difficulty of sorting out the environmental agenda was compounded by the establishment in the late 1960s and early 1970s of a number of new, staff-based, science-litigation-and-lobbying groups such as the Environmental Defense Fund and the Natural Resources Defense Council. At the same time, some of the older conservationist groups, such as the Sierra Club, Wilderness Society and National Wildlife Federation, beefed up their presence in Washington and also sought to *professionalize* by recruiting lawyers, scientists and full-time lobbyists who became the core of their organizations. Though initially adopting an adversarial stance, these groups soon became adept at working the

system of political and economic power, influencing the legislative process, using the courts effectively, and, perhaps most
important, working with industry and government to create an
elaborate pollution-control and management system. In turn,
policy-makers and the media came to situate environmentalism as
an expression of the activities and agendas of these organizations
and the regulatory system they had helped to create.

The development of this relationship between the professionalized environmental organizations and policy-makers was
grounded in part by the interpretation and framing of Earth Day.
Even in the weeks and months prior to the event, a concerted effort was made by both policy-makers and the media to redefine
environmental concern as an issue of national consensus rather
than contention, exemplified by Richard Nixon's 1970 State of the
Union message in which he proposed to "make our peace with nature." While recognizing the breadth and depth of various environmental issues, the thrust of this consensus approach centered
around the idea that environmental improvement would come
about by resolving "technical and mechanical problems that involve processes, flows, things, and the American genius seems to
run that way," as a 1970 *Time* story put it. The elaborate pollution-
control system, subsequently put in place by such myriad laws as
the Clean Air Act, Clean Water Act, Resource Conservation and
Recovery Act and, by the end of the decade, the Superfund legislation, was predicated on the notion that a vast majority wanted
to "clean it up" and that cleaning it up was essentially a matter of
technique. Earth Day launched a revolution, according to this
wisdom, but it was a revolution that put aside the radical and more
challenging question of identifying and restructuring *at their
source* the industrial and urban causes of environmental degradation. This challenge over some of the central features of the systems of production and consumption, such as the rise of the pet-
rochemical industry or the triumph of "disposability" and the loss
of product durability, has more recently reappeared in the form of
the push for "pollution prevention," a kind of catchall term that

has also signaled dissatisfaction with the pollution-control approach.

The identification of environmental analyses and solutions as technical and scientific in nature was magnified when environmental problems were placed in a *global* context, as many such issues were then so defined. Population explosion and catastrophe scenarios, early forecasts about ozone depletion and global warming, and, later, predictions of rapid nonrenewable-resource (especially fossil fuel) depletion, led to a stronger emphasis on *individual rather than social action* on the one hand and a kind of doomsday emphasis on the other. Many of the global problems were real enough, and the focus on them did lead to occasional concrete actions, such as the defunding of the Supersonic Transport (SST) and the banning of aerosol sprays. But the manner of their presentation in the post-Earth Day 1970 framing of environmental issues put the onus of responsibility on the individual consumer, parent or commuter. "It is easy to blame pollution only on the large economic interests," Senator Edmund Muskie wrote in 1970, "but pollution is a by-product of our consumption-oriented society. Each of us must bear his share of the blame. If we want air we can breathe and water we can drink, we must ask ourselves if the extra comfort of the latest technological whim is worth the environmental price."

This logic of individual responsibility led to appeals to have fewer (or no) children or to consume less, positions that some 1960s activists interpreted with alarm as the first steps toward forced sterilization programs in Third World communities at home and abroad or justification for maintaining low standards of living among the poor. Media coverage exacerbated those divisions by continuing to define environmentalism both as an elite movement embodied in the professional groups and as a set of technical problems of global rather than institutional proportions whose solutions depended on individual behavior rather than social change.

As a consequence the wide range of social movements that did emerge in the 1970s and 1980s around environmental issues of daily life and sense of place were barely noticed, whether by policy-makers, the press or even, at times, the professionalized environmental groups, until such time that the protests could no longer be ignored. For example, the energy crisis of the 1970s spurred talk by policy-makers about high-tech solutions such as nuclear fusion or synthetic fuels on the one hand, and appeals for conservation as a form of moral sacrifice on the other. But at the same time, a powerful grass roots movement developed around skyrocketing utility rates, a movement that essentially reformed utility practices throughout the United States and which ulti- mately had far-reaching consequences in the energy arena. In this same period, neighborhood movements mounted widespread op- position to the siting of nuclear power plants. Often led by women, these neighborhood organizations, as well as 1960s-style direct- action protest groups, produced some of the largest environmen- tal confrontations of the time.

This contrasted with the professional environmental groups, which were generally late in focusing on this issue and who relied almost exclusively on the tactics of lobbying, litigation and politi- cal compromise. But it was the neighborhood anti-nuclear power movement that eventually had the most direct impact on the in- dustry, playing a key role in adding to the costs of nuclear plant construction (and the eventual decline to zero of new plant orders) while providing support for alternative energy technologies and conservation strategies. The media were also late in identifying and covering these citizen movements and, in the process, never caught sight of the community-centered, populist form of envi- ronmentalism they represented.

This oversight was especially magnified with the media's fail- ure to identify the rise of the many anti-toxics movements of the 1980s, the most important expression of grass roots/populist en- vironmentalism during this past decade. The rapid proliferation of these neighborhood-based protest groups, many of them ini- tially formed to oppose a particular facility or disposal site, were invariably denigrated by their waste industry opponents as expressions of NIMBY (not in my backyard) politics: selfish, parochial, unwilling to accommodate the public good. This characterization of the NIMBYs was reinforced by media coverage, which focused on the technical nature of the debate over such facilities and the "big" picture about the waste crisis. The so-called NIMBYs, meantime, while out of the media spotlight, nevertheless succeeded both in mobilizing their communities and challenging the assumptions and programs of the waste industry and its public official allies. As a consequence, the neighborhood groups, linked in new networks, essentially changed the terms of waste policy, shifting the focus from disposal and pollution control to reduction, reuse and recycling. As with the nuclear power situation before it, the costs of public opposition added an economic as well as political dimension to the pollution control/waste disposal approach that had not been anticipated by industry, government, the press or the professionalized environmental organizations.

By the end of the 1980s it had become clear that environmentalism meant many different things to different groups and movements. Moreover, questions of gender, race and class also highlighted such differences. The grass roots groups, for one, continued to be defined by the central leadership role that women played in them. Often this reflected the major preoccupation with family, neighborhood and place that issues such as the hazards of nuclear power or toxics touched upon. This crucial female role within the movement was further highlighted by two striking gender-related phenomena. For one, the leadership of women sometimes engendered family crises in terms of husbands becoming unsettled and unhappy with the new prominence of their wives. Secondly, industry and government opponents often questioned the capacity of these women to understand the issues involved. Expertise, in turn, came to be presented as a form of male authority, with the neighborhood groups not only rejecting this use of expertise but arguing that questions of the risks, costs and appropriateness of facilities required community input. The gender question in environmentalism had direct bearing on the movement for environmental democracy.

For the grass roots movements, race also came to be seen as a significant factor, given, for example, the increasing tendency to site waste facilities in poor, minority neighborhoods. Partly a reflection of the increasing land and transportation costs (which made it more difficult to locate facilities outside town limits) and the political conclusion of the waste industry and public officials that poor people of color would be less capable of mounting NIMBY-style challenges, it became more transparent, during the 1980s, that certain environmental burdens also had a racial component. Unlike the professionalized environmental groups, the community-based movements expanded their base to include both multiracial coalitions and minority-led groups, providing a kind of civil rights dimension to this brand of environmentalism. The call for "environmental justice," a central thrust of the grass roots groups, also became a crucial distinguishing feature between the different environmental movements.

Class issues provided another kind of dividing line. During the 1970s and 1980s, the professionalized environmental groups became particularly vulnerable to the industry-generated argument that environmental regulation was synonymous with job loss, a position reinforced by the media's continuing tendency to frame their coverage of specific environment-oriented industrial conflicts along those lines. Though a few attempts were made to create high-level labor/environmental coalitions, a basic mistrust between the professionalized environmental groups and trade unions, some of it couched in class terms, prevented any sustained ties from developing. The one key issue providing a possible connection between the two parties was the question of worker health and safety, and while this issue came to be a substantive concern for certain unions and union-related organizations such as the COSH groups (Committees on Occupational Safety and Health), it remained largely divorced from overall environmental agendas. Moreover the rapid increase in the generation of *toxics*, a crucial feature of the production system since World War II, was once again framed by policy-makers, professionalized environmental groups and the media as a *waste-disposal* and *pollution-control* issue rather than a *toxics-use* issue that required a reduction or pollution-prevention strategy in both the community (in terms of discharges) *and* the workplace (in terms of worker exposures).

The grass roots movements, however, had a greater capacity to develop an approach that incorporated both community and workplace issues. In many instances community residents included those who worked for the particular industry whose emissions or discharges were being challenged. Even where this overlap did not exist, many of the community groups had themselves been forced to confront various forms of job or economic blackmail (you build this waste facility, industry would frequently argue, and we'll provide a certain number of jobs for local residents) and were sympathetic and knowledgeable about the problem. Most important, the willingness and readiness to take on a common corporate antagonist had begun to lead to impressive new coalitions of insurgent workplace and community-based movements, especially with the toxics issue. One key example of this connection—an example that, parenthetically, also failed to generate significant media coverage—was the successful Louisiana-based campaign against BASF, a German multinational chemical company that has both an anti-union approach and a poor environmental track record. The Oil, Chemical and Atomic Workers (OCAW) local that spearheaded the campaign and enlisted the support of a broad range of grass roots groups realized that its most effective message was to identify BASF as an egregious polluter rather than simply as a harsh employer.

The growing significance of gender, race and class among the grass roots movements has paralleled the emergence of the twin demands for environmental justice and environmental democracy, approaches most pronounced in the anti-toxics battles. The question of environmental justice has emerged out of the different ways in which pollution and hazards affect people in communities or in the workplace, which in turn has provided these movements with that "civil rights" flavor found in other "social justice" movements. The fight for environmental democracy, often framed in terms of whether experts or ordinary citizens should make the decisions about whether or how to pollute and what constitutes acceptable risk, has raised larger issues about the political process and its failure to confront the polluters. These demands, in turn, suggest the synthesis of a new and dynamic environmentalism, reminiscent in part of the beginnings of the civil rights and student movements of the early 1960s. Though in conflict at times with the professionalized environmental organizations, grass roots groups as diverse as the largely black and rural community organization fighting the Emelle, Alabama, hazardous waste landfill, the largest of its kind in the country; the "downwind" residents of the nuclear waste facilities in Hanford, Washington; the housewife-led coalition against a medical waste incinerator in the small city of Kenosha, Wisconsin; and the integrated coalition fighting a polluting Chevron plant in a working-class suburb in the San Francisco Bay area have all begun to have an impact on the nature of environmental politics. That impact includes a significant influence on a number of the staff-based organizations, ranging from the large and internationally oriented Greenpeace to the regionally centered Citizens for a Better Environment. Moreover, the stirrings of a new student environmental movement present an additional element in the context of the continuing transformation of a complex and politically and culturally diverse environmentalism.

A portrait of this multiple and often disputatious set of movements has been largely absent from media coverage. To the media, *environmentalism* has continued to refer to the technical maze of pollution-oriented and nature-related man-made problems, a number of them of global scale, which are in turn, it is suggested, tied to a myriad of individual behavior or "lifestyle" issues. Unlike the 1960s movements that sought to expand the dimensions of politics by claiming the "personal is political," this presentation of lifestyle environmentalism, heralded in countless books and tracts and proclaimed in numerous media outlets, has reversed the equation and identified "the political as personal," a fundamental depoliticizing and decontextualizing of environmental issues. This reverse equation, already pronounced in Earth Day 1970, achieved its fullest expression with the coverage preceding and including Earth Day 1990. "You too can save the planet," the media continuously implored, offering tips on recycling, conserving water or energy, and making more (presumably) environmentally sensitive purchasing decisions. Environmental activity became another form of consumption.

THUS WE ENTER the 1990s witnessing the unfolding of two interpretations of environmentalism: a media-framed, technically oriented, behavior-defined set of global issues and individual activities; and a less visible, more contentious grouping of rooted movements, networks and actions. While policy-makers, the media and even industry antagonists have granted semiofficial status to the professional environmental groups, and often turn to them for the "environmental" point of view, this primary interpretation of environmentalism has largely ignored the growing, more populist undercurrent of the movement. At the same time, there has been a failure to distinguish between the professional groups, where differences of approach do occur, and even more significantly, to see how the dynamism of the less visible movements has affected these groups.

For one, the quest for environmental justice and environmental democracy has begun to permeate the overall discourse of environmentalism itself, elevating such concepts as pollution prevention, industrial restructuring, community right-to-know and environmental equity. Instead of simply responding to issues of the moment (based perhaps on the latest scientific report given the most extensive media coverage), a debate has begun within the broad confines of environmentalism over whether and how to achieve that more populist cast and create agendas more closely responding to the concerns of daily life. From the earlier admonition of "think globally, act locally" has come a contemporary version of both "think locally, act locally" *and* "think globally, act globally."

Global issues, indeed, are increasingly understood to be grounded in daily life, with the debate focused on how to translate that connection into a new politics and, beyond that, into a new and more democratic discourse capable of challenging the prevailing wisdom and institutions of our contemporary industrial and urban order. The environmentalism of the future, depending on the outcome of such debates, offers the potential of generating new kinds of coalitions around the conditions of daily life. And it is out of that debate that there can also emerge the quest to create a more democratic, just and globally secure society where the question of environment is central to the decisions that govern different systems of production, the social relations they engender, and the ability, as Barry Commoner has put it, to make our peace with a battered and seriously degraded planet.

EVERETTE E. DENNIS

In Context: Environmentalism in the System of News

Any deficiency in public understanding of the current state and probable future of the environment could be solved by continuous and massive press coverage. Many environmentalists and environmental experts—whatever their point of view on other matters—believe this. So do environmental journalists and other communicators, whether they adhere to the cant of objectivity or prefer advocacy journalism.

If it is all so simple, why don't the news media simply develop a strategic plan and make it happen? The answer is simpler still: The environment, while profoundly important, is seen by the media as one of many topics worthy of coverage, but at the same time one that is all too often what *The Economist* calls "a snore." What the various news media decide to cover is at one level a daily bargaining game between and among competing interests and topics. That bargaining game is guided partly by tradition, but also by structural patterns within the press that value certain subjects and situations more than others.

The wanderer: Loaded with solid waste from New York State, the garbage barge journeyed the world's oceans in 1987 looking for a port that would accept it.
Courtesy Associated Press

The media, of course, have much more to do than cover the environment. And with many barriers to regular and effective coverage linked to reasonable coverage of other areas, the difficulty deepens. Some critics believe that the American people (as represented by the media audience, which is not by any means all of the people) can handle only three topics at any one time. That means that if the Persian Gulf war is on the agenda, there might be room for the Soviet Union and for the stock market, but probably not much room for domestic problems, China, Central America and the developing world. This is a real problem for anyone or any interest hoping for continuity of coverage of topics that are not always riveting.

The best and most thoughtful of editors and broadcast producers try to balance what they regard as genuinely important and newsworthy with what the public will want to watch and read about. This is a matter of both substance and style. Any look at the content of the news media, either short-term or over time, informs us that international affairs, politics, economics, crime, celebrity, sports, science and a few dozen other topics dominate the content of the media. Although the contours of these topics change over time, with the news choices of one generation differing in degree from those of the next, it is clear that there are many traditional news commitments.

Students of newsmaking offer instructive guidance on the factors that make news in America (standards differ in different countries and societies, although there are a few universal themes). Along with a co-author I tried some years ago to find an adequate definition of news to guide a discussion such as this one on news and the environment. This is what we came up with:

• *News is a report that presents a contemporary view of reality with regard to a specific issue, event or process. It usually monitors change that is important to individuals or society and puts that change in the context of what is common or characteristic. It is shaped by a consensus about what will interest the audience and by constraints from within and out-*

side the organization. It is the result of a daily bargaining game within the news organization that sorts out the observed human events of a particular time period to create a very perishable product. News is the imperfect result of hurried decisions made under pressure.

• *News is an organizational product and is guided by news values; among these are impact on people and society; timeliness; prominence of those involved; proximity to the readers and viewers; the bizarre, conflict, and others.*

Within this general framework, news organizations have editors responsible for "covering the territory," which is defined as the whole world in a broad sense, but especially the nation, the state and the community. Beyond that, there are specialists who take up politics, economics and other topics deemed important and vital. Just as all news is not deemed equal, neither are reporters equal colleagues operating under a code of fairness and equity. For reasons of tradition, people who cover politics are thought to be more important than people who cover education, religion or the environment. This means they have higher status, get better assignments, and more often see their work designed as a major story—on the front page or at the top of the newscast.

Not long ago I was with two editors who were talking about the relatively new Society of Environmental Journalists. They wanted to know whether this was just another advocacy group, a self-serving interest, seeking favor and positive coverage in the media. When they were told that the group was more like the National Association of Science Writers (NASW), eager to have their speciality get its due, one of the editors said, "But I've heard that some of them are really advocates for the environment." While many citizens would say, "Yes, who isn't?" to many journalists this is anathema. The idea that a reporter would openly espouse a point of view on a public issue violates the accepted canons of journalistic practice. While many in the public distrust the media and see their interests as profitability and sensationalism rather than public service and effective communication, the press sees itself as an instrument of public communication that "separates fact from opinion" and tries to present its audiences with impartial reports.

The American media's ideological code of objectivity describes both a philosophical commitment to what media people call "fairness" and a style of news presentation that differs greatly from the textured interpretive essays of European journals. Beyond the separation of fact and opinion (ostensibly distinguishing news stories from editorials, columns and commentary), traditional journalists in America strive for emotional detachment—an essential element of their professionalism, they say—as well as balance—typically stated as giving "both sides" or conflicting interests a shot at a level playing field. Of course, these formulations are simplistic and there are volumes of learned press criticism that attack them. Still, they are the prevailing beliefs, attitudes and standards that guide the news media, and the fact is that any interest must live and deal with them if it is to get any attention at all in our media system.

INSTITUTIONALLY, THE MEDIA and environmentalists see each other in quite different ways. To many editors, the environment is one of many important and compelling topics the public needs to know about. Although the Earth, the waters and the sky have been around far longer than the media, the environment has never been a comfortable arena of coverage for the media to handle. At an abstract level, it cuts across all news beats and topics. The environment may be a political-government topic. It may, and often does, have a great deal to do with the economy and with commerce. It is also a scientific story. Thus some aspect of "the environmental story" may be in the purview of many different reporters and editors. While the topic of conservation—what we used to call the environment before 1970 or so—used to get modest coverage, it never had much priority in the media and was often confined to outdoors coverage in the back of sports sections, the stuff that hunters, fishermen and other traditional "nature-loving" readers might care about.

Moreover while national parks, wildlife, water pollution and other topics occasionally rise to the highest news priority, as with the Exxon *Valdez* oil spill in Alaska or the eruption of Mount St. Helens, nature and the environment have a somewhat eccentric image in American life, going back to early explorers and the "different drummer" approach of Henry David Thoreau.

The structural and professional conservatism of the media maddens many who have a passion for environmental issues and causes who cannot understand anyone who does not regard saving the Earth as the most compelling priority in the world today. Yet it is a minority of news media leaders in America who have given "the environment" their seal of approval as a news story of the very highest priority. While many will say it is, a look at their news choices suggests otherwise. One of the reasons is that the media still value people and ideas in conflict more than conditions and trends that are omnipresent. While a great environmental disaster will attract attention, scientific reports on acid rain or air quality will get only fleeting notice, mainly because the human interest factor is thought to be limited and ephemeral.

Thus a story like the multifaceted nature of environmental change must be filtered through traditional news media beats. And given a chance to write about presidential candidates or environmental hazards, most reporters on the move and on the make will choose the political story anytime. Recently, much has been written about the failure of presidents and presidential candidates to make the environment a matter of continuous public concern. It is, one commentator said, too often a snore, a hard sell in a world where people like scandal and sensationalism, personalities and celebrity. "The environment," he said, "just isn't sexy enough" to push out news that is.

The debate among environmental reporters about advocacy and objectivity is an important one, but also one that will fuel the suspicions of news executives who are always on the lookout both for self-serving special interests and for reporters who have lost their objectivity. As one who believes that objectivity is more an ideological icon in our media than an operational reality, I still recognize how deep the passions on this topic go. In earlier decades, reporters argued that human rights were so important that they could not and should not be impartial about the civil rights struggle. More recently, some newspapers and television stations have fashioned themselves as warriors in the war on drugs and have not only covered the story, but helped develop information campaigns to promote drug-free schools, for example.

However worthy the issue or cause, the goal of marshalling and shaping public opinion causes nervousness among journalists and media owners who think such advocacy could compromise the essential independence and integrity of the press.

WHEN IT COMES to systematically covering "the environmental story," anyone who moves beyond the most simplistic approach sees immediately the extraordinary complexity involved even in mapping the territory, let alone understanding trends, issues, conflicting evidence, the role of information sources, and other aspects of the story. Most environmentalists and environmental scientists I have met are generally unhappy with press and media treatment of a subject they hold dear. And while this is almost always true with others representing various interests and specialities, from education to military affairs, there is a vehemence and extraordinary frustration in the environmental field. There is rightly a feeling that the story is extraordinarily complex, so much so that the media—which hate ambiguity—may not be able to cover the territory any other way than superficially. An essential problem is the lack of expert training and expertise among media people. Reporters can easily be influenced on this topic by special interests or be put off by those they regard as shrill and impatient about news coverage of their most important concern.

While the environmental field sees the media as an occasional ally, they seem more often a fickle and unreliable friend, one whose attention span and commitment to serious, long-term coverage is unreliable. Typically, the environmental spokesperson does not understand the media's conventions and ground rules, let alone their traditions or institutional culture. They believe that the environment ought to be covered extensively and well because it is right to do so. And while this same goal is present in the minds and professional commitments of a few editors and scores of environmental journalists, they are not powerful enough to make it happen.

Others in the press are very suspicious of environmental spokespersons and causes, feeling that while they speak in platitudes that are hard to dispute or disagree with, their real interests and passions are not always up-front. Journalists have experience with environmental news sources who are secretive, arrogant, uncooperative and punitive (when they don't like a story). While this is true in virtually all fields, the environmental crowd sometimes has a quasi-religious zeal about their passions and are impatient with the more measured approach.

A special problem is the anti-business stance of much of the environmental movement, whether real or a matter of appearances. As business and environmental interests are increasingly pitted against each other, this natural contentiousness makes reporters and editors leery about whom to use as a source of reliable news and information. Given the complex nature of environmental sources, ranging from scientists and economists to political activists and even some who use terrorist tactics, the playing field is not only not even but encumbered by furrows and bumps that make it difficult to scope out the players and what they are doing.

IN THE END, the news media really do cover those topics and issues that *interest the public*. As critics have pointed out for a hundred years or more, there may be a yawning chasm between what interests the public and the public interest, but nonetheless only stories that are salient and pertinent to people in their daily lives (or their fantasy lives) have any chance of getting regular and reliable coverage. Thus those who care about seeing quality environment coverage—whether they are involved with the environmental movement, covering the environment for the media, or simply members of the public with a desire to know more about the topic—must grapple with the public attention span. They must, in fact, become expert in persuasive communication, at finding angles and connections between and among stories that will ignite public interest and inspire a desire and demand for more information.

Almost any rational and responsible person recognizes the importance of the environment, as well as the need to have the public well informed on the subject. And by the same token, most knowledgeable observers of the media will probably, in the end, acknowledge that the media alone might not be able to do the job. While there ought to be regular and continuous coverage of the environment, this reporting will always have to compete with other matters of importance in public life.

It is not enough to exhort the media to cover the environment more seriously and systematically. The media's performance is better today than ever before; still, those who truly care about public understanding should also take into account the media's limitations. An informed public will require schools, political parties, religious organizations and other institutions of society to become part of the system of environmental information. This can happen only if systematic efforts are made by environmental interests and news sources to develop alternative information strategies, recognizing that our system of freedom of expression, while reliant on the news media, requires much more to be whole.

Although the environmental fields cannot speak with one voice, they can in individual ways fashion strategic information programs that will reach various constituencies effectively. Much pure information these days reaches the public through data bases and in popular culture, without much involvement by the news media. Those in the environmental movement need to work effectively to develop awareness and sensitivity in the public, to help map out the territory so that the average citizen can know and understand what is at stake, how that is happening and who cares about it. Someone, for instance, must stay in touch with Hollywood and television screenwriters, as well as with opinion leaders of all kinds in labor, education and the church.

The environment is such a compelling concern that it is hard to imagine that there isn't at a fundamental level nearly universal support for it. But such a view is too simple for public policy generally and for the news media specifically, which face an increasingly pluralistic world with hundreds, if not thousands, of competing topics, issues and constituencies to be reckoned with. The environment is one of them, and the sooner environmental interests realize that they are not the only wheel in town, the sooner they can find a strategic approach to communicate effectively through America's diverse media system.

COVERING
the
ENVIRONMENT
as if
IT MATTERED

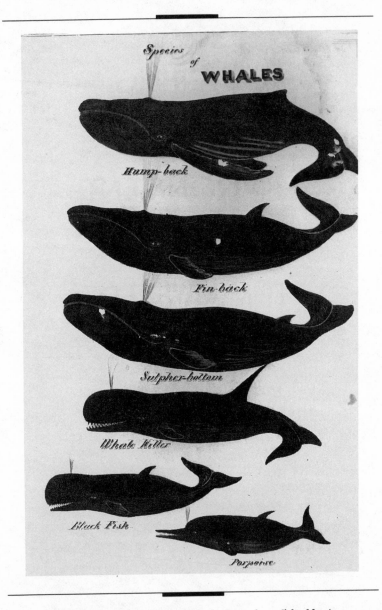

Species of Whales Seen on Board Whaler Lucy Ann: *John Martin,*
Wilmington, 1843
Kendall Whaling Museum, Sharon, Massachusetts

DONELLA H. MEADOWS

Changing the World Through the Informationsphere

———

The lithosphere, the hydrosphere, the atmosphere, the biosphere—rocks, water, air and life—these, the ecology texts say, are the spheres that make up the world. Unfortunately for those who want to do something effective about the environment, that list leaves out the most important sphere for action, the only one over which we have much control. That sphere is called by various authors the noosphere, the sociosphere, the technosphere. I call it the informationsphere, because information is what makes it up.

Ecologists rarely include the human mind and its products within their otherwise holistic field of study, but there are people who do study the sphere of information. They call themselves information theorists or systems analysts. That is what I am, or was, before I left the university to become a newspaper columnist.

From the academic point of view, that move was inexplicable. From my point of view it was a simple shift from theoretical to applied systems analysis. I made the change because I couldn't stand any longer to see what textbooks call the biosphere falling apart. I wanted to do something about that. I had to do it through the media, because systems theory taught me that there, squarely in the informationsphere, was the place to change the world.

This is how systems analysts see the world and the role of information in it. First, we say, the world is made up of "system states." These are for the most part not information, but physical, measurable stocks that constitute whatever system we're interested in: the number of people with AIDS, the amount of carbon dioxide in the atmosphere, the balance in your checking account, the remaining area of tropical forest, the national debt.

System states change because of flows into and out of them: the AIDS infection rate, the AIDS death rate; the release of carbon dioxide from fossil fuel burning; deposits into and withdrawals from your account; the growth, clearing and burning of the forest; the national deficit.

Many flows are natural processes of the litho-, hydro-, atmo- and biospheres. Many others are initiated or governed by human decisions and actions. That's where information comes in. *Information is the signal that tells an actor the state of the system*—the signal that alters the decisions that govern the flows that change the world. The gross national product is information. So are the Dow Jones average, a thermometer reading, a satellite picture of the ozone hole, prices, infant mortality rates, bank statements, news stories, films of a tropical forest burning. All information, leading to decisions and actions.

Information is the vehicle for feedback in systems, and feedback is the means of control. A feedback loop is a circular chain of causation from the physical world (systems states and flows) to points of decision and back through action to the physical world. Half of every feedback loop is information. In that half the system state is compared with some goal. If there is a discrepancy between what is and what someone thinks ought to be, that brings about a decision. The decision may be influenced by perceived constraints, resources, costs, benefits, other goals and discrepancies—all information. If that information leads to an effective decision, there is an action in the physical part of the feedback loop, which brings the system state closer to the goal.

In real systems there are many intersecting feedback loops. Each actor monitors many different system states, and each system state is influenced by many actors. The whole business is held together by information.

For a feedback loop to bring the system state to the goal, its information must be timely, accurate and noticed. If it isn't, if information is delayed, biased, noisy, sequestered, ignored, denied, distorted or absent, the system of which it is a part *simply can't work*. You can't solve a problem that you don't know about, or that you find out about too late, or that you have exaggerated or underestimated. You can't reach a goal you haven't defined. You can't make intelligent choices unless you are informed about those choices and unless you have (and are open to) accurate information about how past choices turned out.

Systems analysts are attuned to all the travesties brought about by people acting upon faulty information. Therefore we tend to develop strong moral feelings about the informationsphere. We have an Eleventh Commandment: Thou shalt not distort, delay or sequester information. (This commandment must be even more frequently breached than the other 10.) We are always trying to clean up information streams or to supply missing information to decision points. We think that the reordering of information, rather than the shoving around of things or people, is the most powerful way to change the world.

For example, I like to tell the story (a true one) about a Dutch residential development full of identical houses, except that by accident in some houses the electric meters were installed in the basement, while in others the meters were put in the front entry halls. Where the meters are in the halls, electricity use is 30 percent less than where the meters are in the basement. To the people who live there, the daily sight of those meters whizzing around provides information that changes their actions. No price differences, taxes, rules or moral pleas. No coercion. Just information.

From a systems point of view the way to clean up the nation's water is to require each town or industry to put its intake pipes *downstream* from its output pipes. That way each user gets instant feedback on the quality of its effluent. The water would get cleaned up and the Environmental Protection Agency wouldn't have to do anything but inspect the location of pipes. So far no one has tried to implement that policy.

Another example: It is a grave mistake in systems terms for governments to rate their economic performance by the growth rate of the gross national product (GNP). If ever-greater GNP is the goal of societal feedback loops, the system will contort itself trying to produce ever-greater GNP, which is environmentally impossible and not even what anyone wants. GNP is a measure of economic activity, of throughput. It is "the fever chart of our consumption," as Wendell Berry says. It is cost, not benefit; quantity, not quality or justice or security or welfare. It isn't even wealth. Wealth is the stock of goods, not their flow. A society would be as well off and environmentally more sustainable with a stock of houses, cars, factories that is replaced slowly rather than quickly—though it would have a much lower GNP.

Be sure the goals of your feedback loops are your real goals, we systems analysts say, when you're trying to fix a messed-up system. Take information from the truly important system states. Deliver it accurately and promptly to decision makers. Make sure they notice and understand it. Otherwise the behavior of the system will be deranged.

Making systems manageable is one reason to act through the informationsphere and to care about its integrity. For a long time I thought it was the only reason. Then I wrote a book called *The Limits to Growth*, which was intended as a small piece of information, a signal to warn about the environmental consequences of orienting social feedback loops around the goal of limitless growth. The response to that book showed me something new about the informationsphere. The signal I sent out was not received. It was *actively* not received. There was a *strong commitment not to receive it.*

That's when I learned about paradigms.

THE PARADIGM OF a society (to expand upon the theory of Thomas Kuhn in *The Structure of Scientific Revolutions*) is its deepest shared beliefs, its unspoken, almost unconscious commitment to how the world is.

In societies that speak Indo-European languages, for example, the use of nouns and verbs is a paradigm that tells us the world is made up of *things* and *processes*. In some other languages there are only processes.

The Western paradigm draws sharp distinctions between the body and the mind, between work and leisure, between the economy and the environment. In other paradigms, these divisions make no sense at all.

People are profoundly uncomfortable in the presence of paradigms not their own. If you are pro-life talking to someone who is pro-choice, or a capitalist talking to a committed Marxist, or an ecologist talking to an economist, or vice versa, you know what I mean. Something about those conversations makes you feel not only sputteringly angry, but slightly dizzy, as if the earth were shaking under your feet. The earth isn't shaking, but your world is—your world as your paradigm defines it.

If you're like most people, you don't hang out with folks who shake your world. To keep yourself on solid ground, you think of ways of dismissing, demeaning and avoiding them. (That's what happened to me when I challenged the paradigm of endless growth.) Together with people who share your world view, you form a culture, and out of that culture you do your best to build the world as your paradigm tells you it is and ought to be.

Which is to say, *widely shared paradigms bring forth systems.* The paradigm of the ancient Egyptians brought forth papyrus and pyramids; that of Hapsburg Vienna brought forth gilt cherubs, opera houses and Sacher torte; the modern industrial paradigm brings forth skyscrapers and acid rain, personal computers and global warming, heart transplants, oil spills, nuclear power, stock market booms and busts.

Our paradigm tells us what system states to care about, what goals to set, what signals to monitor. A change at the level of paradigm first figuratively and then literally restructures the world. If you don't believe that, just sit back and watch the effects of one paradigmatic change—*glasnost*, the simple opening of information flows—on Eastern Europe.

Ralph Waldo Emerson saw more than 150 years ago how paradigms bring forth new physical realities:

> Every nation and every man instantly surround themselves with a material apparatus which exactly corresponds to their moral state, or their state of thought. Observe how every truth and every error, each a thought of some man's mind, clothes itself with societies, houses, cities, language, ceremonies, newspapers. Observe the ideas of the present day . . . ; see how each of these abstractions has embodied itself in an imposing apparatus in the community, and how timber, brick, lime, and stone have flown into convenient shape, obedient to the master idea reigning in the minds of many persons. . . .
>
> It follows, of course, that the least change in the man will change his circumstances; the least enlargement of ideas, the least mitigation of his feelings in respect to other men . . . would cause the most striking changes of external things; the tents would be struck; the men-of-war would rot ashore; the arms rust; the cannon would become streetposts.

A society that refuses to consider the idea that there are limits to growth is not going to bring forth a physical economy that fits within the constraints of the planet. A society that thinks there's an "away" to throw things to is going to find itself choking on its own waste. People who do not see nature as the support base for all life, including their own, will destroy nature and eventually themselves.

That's why 5 years ago I resigned my professorship to become a columnist. I knew I had to do more than clean up a few information flows. I had to challenge some central tenets of my society's paradigm. Here are just a few of the common ideas I set out to question:

• One cause produces one effect. There must be a single cause of acid rain or cancer or the greenhouse effect, and we just need to discover and remove it.

• Improvements come through better technology, not through better humanity.

• The future is to be predicted, not chosen or created. It happens to us; we do not shape it.

• A problem does not exist or is not serious until it can be measured. Quantity is more important than quality.

• Economics is the measure of feasibility. Current economic reckoning of costs and benefits is complete and correct. If something is economic, it needs no further justification. If it is uneconomic, it can't be done. (As E. F. Schumacher said, "Call a thing immoral or ugly, soul-destroying or a degradation of man, a peril to the peace of the world or to the well-being of future generations; as long as you have not shown it to be 'uneconomic,' you have not really questioned its right to exist, grow and prosper.")

• Relationships are linear, nondelayed and continuous; if we've done something and gotten a certain effect, then we can do twice as much and get twice the effect. There are no critical thresholds, and no surprises.

• Feedback is accurate and timely; systems are manageable through charging ahead and wiping up problems as they occur.

• You can measure results by effort expended—if you have spent more for weapons, you have more security; if you use more electricity, you are better off; if you spend more for schools, your children are better educated.

• Nations are disconnected from each other; people are disconnected from nature; economic sectors can be maximized independently from each other; some parts of a system can thrive while other parts suffer.

- Choices are either/or, not both/and.
- Possession of *things* is the source of happiness.
- Individuals cannot make any difference. (Unless they are media-anointed superstars, in which case they can make any kind of difference overnight. Gorbachev should have been able to transform the entire Soviet economy by now.)
- The rational powers of human beings are superior to their intuitive or moral powers.
- Present systems are tolerable and will not get much worse; alternative systems cannot help but be worse than the ones we've got.
- We know what we are doing.

I would submit that these widely accepted assumptions can lead and are leading to terrible consequences. The list is not complete, but it's enough to keep a column going for a long time.

A PARADIGM IS upheld by the constant repetition of ideas that fit within it. It is affirmed by every information exchange, in families, churches, literature, music, workplaces, shopping places, daily chats on the street. The key to paradigm stability and coherence is repetition. Therefore when people learned how to repeat information on a mass basis—to make printing presses and send messages over electromagnetic waves—they not only created tools with the potential to improve vastly the information flows in systems, they also inadvertently invented potent techniques for paradigm affirmation and, theoretically, for paradigm change.

Every advertiser, world reformer and power broker knows that. It's no accident that the Reverend Moon has acquired his own newspaper, that the Audubon Society makes nature films, and that the Soviet Union in its central planning days saw no need to bring indoor plumbing to distant villages but took pains to bestow electricity and color TV sets.

Though the media can be instruments of social change, they

seldom are. They even fall far short of their potential to improve information flow. In the Soviet Union the clumsy propaganda they carried was so incongruent with everyday experience that it became not information but self-parody. Western media carry much more skillful propaganda, but its purpose is to entertain, lull and sell, not to inform, much less raise questions about paradigms.

Because of that purpose, because of the structure of their ownership, governance and incentives, and because of the larger paradigm within which they operate, the Western media have developed a series of problems well known to most everyone:

• They are event oriented and shallow; they do not report underlying structure, historical context or long-term implications.

• Their attention span is short; they create their own fads, follow them and tire of them; they seem to discover the environment as an important subject only every 20 years or so.

• They are attracted to personality and authority (which they have themselves created); they are uninterested in people they've never heard of.

• They simplify issues; they have little tolerance for uncertainty, ambiguity or complexity.

• They love conflict and controversy and find harmony boring; they portray the world as a set of win/lose, right/wrong situations.

• They tend to force the evidence to conform to their story, rather than to see the world as it is. (There is no more frustrating experience than to try to convey to a reporter information that directly contradicts his or her "story.")

• They operate from skepticism; they have been manipulated so often that they don't believe anyone; therefore they convey a lack of trust in all information.

• They report through unconscious filters of helplessness, hopelessness, cynicism, passivity and acceptance. They describe problems more than solutions, obstacles more than opportunities. They systematically unempower themselves and their audience.

Above all they are deeply conservative, even those that try to be "liberal." They uncritically transmit, reflect and amplify the reigning societal paradigm. Newscasters report a rise in GNP as if it were both important and good. We hear about industries "creating jobs" for workers, not about workers creating profits for industry. The environment according to the media is a luxury, something beautiful but trivial, not something that supports our lives.

Editors across the nation topped stories of the logging controversy in the Pacific Northwest with headlines such as "An Owl Versus an Industry"—which tacitly accepted industry's exaggerated view of the situation. The industry is not threatened in the least. The owl is. The old-growth forest is—an entire ecosystem with thousands of species. Jobs are threatened, but by the forest companies' own labor-saving technologies, by their export policies and by their unsustainable pace of harvesting, not by the owl. The headline could have read more accurately "A Forest Versus Greed." But it's not quite permissible to say that, not in headline-size letters.

Commercial pressure is partly responsible for cop-outs like that, but more responsible is the paradigm that pervades everything in the culture, including the media—even the hallowed halls of public TV. I lost a paradigm fight over the title of a PBS environmental series that was tentatively called "State of the World." That was an inoffensive title in any paradigm, but a boring one. Surveys showed that a lot more people would tune in if the series were called "Race to Save the Planet."

But that's inaccurate and arrogant, I argued. The *planet* isn't in trouble, and if it were, human beings couldn't save it. That's just the kind of ludicrous self-aggrandizement that is creating the very environmental problems shown on the programs. How about a title that puts us properly in our place, one that admits that what's endangered is *us*, or at least our wasteful way of doing things? How about removing that unnecessary focus on doom (consistent with the present paradigm, in which any change is doom) and turning, as the programs themselves do, toward a positive image of the sustainable world we have a historic opportunity to create?

Even at PBS ratings are more important than accuracy, and surveys are the source of wisdom. "Race to Save the Planet" it is. All-powerful humanity rides in on a white horse to rescue fragile, trembling Planet Earth.

That decision was made by good people caught in a bad system. It's a system trapped in what my colleagues call "drift to low performance." The trouble is that when a society's *information* system drifts to low performance, the society cannot correct or govern itself. It can't deliver essential feedback. It can't identify the places where its paradigm leads it astray. It can't talk about its real problems. It can't seize the opportunities in front of it. It can't make the right choices, especially not if it is a democracy.

Why try to communicate messages of complexity, of long-term thinking, of connectedness, empowerment and paradigm change through a powerful, resistant and eroding information system? Well, the most optimistic answer is that I think it can be done. A more certain answer is that there's no other choice. The most pessimistic answer is that if we don't, and therefore our society does not seize its opportunities and make the right choices, our industrial civilization will destroy itself and take a lot of innocent species along with it.

The planet, however, will go on orbiting serenely around the sun. The cyanobacteria will probably survive and patiently begin evolving new forms of life.

WELL, AS I said a while ago, the future is not foreordained, it's a choice. As individual human beings, we have the power every time we launch a single word into the informationsphere either to join the murmur of mindless repetition of old, unworkable paradigms, or to affirm new ones. For the moment we have to take the physical and institutional structures of the present informationsphere as given. But the content we put through them can work change at the deepest level—the level of paradigms out of which whole systems arise.

That's what I try to do, in simple, humble ways. I don't write about "growth" with unquestioning approval. I ask what the growth is for, who will benefit, at what cost, paid by whom. I avoid phrases like "create jobs" and "save the planet." I don't admit to false dilemmas between economic growth and environmental quality. I don't apologize for my belief—actually my moral certainty—that everything in nature has its own value, a far greater value than we humans know how to measure.

I try not to say "global warming *will*. . . ." Global warming *would* or *might*, if we let it happen. Only a small amount of warming is now inevitable; more than that is still a choice.

Though I go out of my way to present opposing views, I don't confuse willful ignorance with real uncertainty, and I don't equate the viewpoint of a perpetrator with that of a victim. The voice of the Plum Creek Timber Company just is not credible when it comes to old-growth forest harvesting rates. The nuclear power industry has earned no points for accuracy and many for deliberate deception. The pesticide industry is not an unbiased observer on pesticide safety. The people who defend the spotted owls, the safety of their neighborhoods and the health of their children are being selfish too, in a way, but theirs is a broader selfishness, speaking for much larger community interests than profit-making. They deserve more than equal media respect, space and time.

I present possible changes as if they were changes, not sacrifices. I do my best to be honest about what I don't know and what no one knows. (Think how many expensive mistakes we would avoid if we would just admit that most public policy is based on darkest ignorance.)

Above all I try not to let myself dump on peoples' visions of a better world. I don't demean talk about ending hunger, about powering the world with solar energy, about getting rid of all nuclear weapons. How do I know? If the Berlin Wall can fall, anything can happen. My instinct to distance myself from visionary talk comes partly from my culture, which for no good reason prefers the bias of cynicism to that of idealism, and partly from my own deep inner hurt at all the ways the world is less than ideal. Those promptings from my culture and emotions are not a worthy base from which to shoot down the noblest dreams of my fellow citizens—which are my dreams too.

I do my best to remember that the purpose of my writing is to search for truth and to empower others to do the same. It is not to judge, accuse or rob anyone of dignity or self-respect. Every one of my readers, I keep telling myself, can be a key to the workability of the world.

None of that is easy. I'm a child of my culture too. I have to sell columns to newspapers that have to sell ads that have to sell, among other things, nuclear power and pesticides. Even if those commercial pressures were not upon us all, the truth is too slippery to keep firmly in hand. No paradigm is true, not mine, not yours. But we know much more than what we are, as a culture and society, currently saying.

We know that our media-based informationsphere is not serving us and must be restructured in major ways. We know that there are limits to how much material and energy we can mobilize on a finite planet. We know that once our basic needs are met, more material and energy are not actually what we need or desire. We know that the changes we have to make to live together in harmony with each other and the earth are great changes indeed—but not sacrifices. And though we don't know how much time we have to make those changes, we are pretty sure that time is short.

Said Buckminster Fuller: "All of humanity is in peril of extinction if each one of us does not dare, now and henceforth, always to tell only the truth and all the truth, and to do so promptly—right now."

Oil slick from the tanker Torrey Canyon *on the Cornish coast of England,* *1967* *John Reader/* Life *magazine © Time Warner Inc.*

TEYA RYAN

Network Earth: Advocacy, Journalism and the Environment

Let me first submit to you that, with respect to the environment, advocacy journalism is a misnomer. Think a minute: Who do you know who is against the environment? We can argue about the degree to which we should protect it, or we can argue about the ways in which to protect it, but we're not likely to find anyone who thinks we shouldn't make any effort to protect it. In a sense, therefore, you can't take an advocacy position on the environment.

There are people who will say that you can't even put those two words—advocacy and journalism—together, that they constitute an oxymoron, and anyone who has made a career as a journalist is familiar with the ethic behind that argument and respects it.

The creed of a "balanced perspective" has been the hallmark of journalism for the past 30 years. But I wonder if it isn't time in the 1990s for a different kind of reporting, a different kind of presentation to the public, one that says simply: "This is what I saw as a reporter. This is who I talked to. This is my perspective, and here are my suggestions for change. If you want another point of view, find it from another broadcaster or newspaper."

Most television reports about the environment—news or features—provide dual perspectives, opposing views presented as equals almost as if they were intended to cancel each other out. As a viewer I have often thought, "That was informative, but it doesn't take me anywhere. It doesn't advance me any further. Where do I go now? What do I do with that information?" As historian Christopher Lasch has argued in the *Gannett Center Journal*, what people need is not simply *more* information, but *usable* information that allows them to engage one another as citizens. Unless it is responsibly done and extremely well done, the dueling perspectives approach, I believe, creates apathy and does not empower people. And empowering people is, in part, what covering the environment is all about.

I BEGAN MY journalism career as a reporter for the *Vancouver Sun*, and, along with the rest of the post-Watergate generation, I was dedicated to "uncovering the underbelly—finding the investigative truth." Perspective, analysis and opinion were not part of the beat. I eventually moved on to public television, at KCET-TV in Los Angeles, where I spent 6 years producing documentaries, some of them with environmental themes, though not all of them. The forum had changed—from print to television—but not the rules. A "balanced" presentation was still the standard we aspired to.

But as the air in Los Angeles grew browner, more debilitating; as reports of massive dolphin slaughters made headlines; as the destruction of the world's rainforests became widely known; and as everyone wondered where to put the trash—particularly the plutonium—I wondered if "balanced" reporting was still appropriate.

While the Public Broadcasting System has done a stunning job over the last 15 years of creating and broadcasting programs on environmental issues, there is a tendency for outlets such as PBS to speak to the converted—this often has great impact—but there is a large audience that is left out. I felt the realities of global environmental destruction needed to reach a broader, more mainstream audience, an audience that is not likely to watch an hour-long, very sophisticated discussion of the issues. There has been no voice for a more mainstream network audience, no voice that focused on solutions, no voice that was empowering.

This voice was finally developed at Turner Broadcasting. Certainly a leader in producing and broadcasting programming on environmental issues, Turner seemed to be the logical home for a weekly series on the environment.

As of Aug. 12, 1990, TBS SuperStation began airing a weekly half-hour series, "Network Earth." The program is produced at CNN—by a production staff with a strong journalistic background—but, because of its commitment to advocacy, the show airs on TBS.

As the senior producer of "Network Earth," I know that in order to reach a broad, mainstream audience, our first responsibility is to entertain; our audience might listen if this particular program on the environment doesn't seem to be a bitter pill. Within the fast-paced magazine format there is room for not only reports on serious issues—such as the impact of war on the environment, or the realities of ecoracism—but also interviews with celebrities involved in environmental issues and profiles of "local heroes"— everyday people who are taking steps to protect their corner of the world.

Perhaps most important, "Network Earth" is a program that focuses on solutions. We let our audience know there are answers, that there are ways they can influence the destiny of life on the Earth. Our most distinctive response to this search for solutions was to create a national computer forum through CompuServe, which, with 600,000 members, is one of the largest public-access computer services. Our viewers can dial into the "Network Earth" Computer Forum to find out more information on issues discussed in the program and to participate in weekly conferences with the "Network Earth" staff or experts. For example, we intend to organize computer conferences between students across the nation and environmental experts. We encourage our audience to communicate with the "Network Earth" staff, to up-load story ideas and to help us create a new form of participatory television.

Entertaining. Participatory. "Network Earth" goes to the heart of advocacy journalism. It is a program with a point of view, a program that will always come out on the side of the environment. And, perhaps, it is a program that should stand alone.

ADVOCACY JOURNALISM IS most responsible and works best if not everyone does it (and certainly many news media do not want to do it). Some organization—a CNN, *Washington Post*, *New York Times* or *Los Angeles Times*—has to provide what would be considered a more balanced perspective, the kind that, in the midst of the Exxon oil spill, for example, provides the oil company's perspective as well as the fishermen's.

Other programs do that, and they do it extremely well. The morning news programs on all three broadcast networks do a good job covering environmental issues, as does the "ABC Evening News" with its "American Agenda." The *Los Angeles Times* over the two years 1989 to 1990 has done an excellent job of keeping environmental issues on the front page. CNN, certainly, has made a strong commitment to covering the environment as well. Without this kind of counterbalance, advocacy may not have a place. But once this exists, then I think the environment may be the one area where you can say advocacy journalism is appropriate, indeed, vital.

This is not to demean other coverage—presidential elections, for example—where you could—and indeed some do—make a similar argument. But we're not talking about the Pledge of Allegiance here; we're talking about *survival* in the grandest sense of the word. Moreover the environment may be the one area where the issues are so crucial and so complex that it is imperative reporters offer the public some guidance.

There is an argument that all environmental reporting, balanced or not, is advocacy, just because it raises awareness of these issues. That was certainly the effect (intended or not) of the *Time* cover of Earth as "Planet of the Year" in 1988; suddenly the environment moved from being a fringe issue, largely forgotten over the past decade, to an issue of paramount concern to the public.

"Network Earth" takes this concept a step further: It points the audience in a certain direction. I am not an advocate of fist-waving, but, given the gravity of the issues, I do think it is time for a wake-up call. And that is what we do.

An example: In our premiere program, we traveled to Panama to investigate the effects of the U.S. embargo and the subsequent invasion on the Panamanian forests and the Panama Canal. American efforts to oust Manuel Noriega left many Panamanians out of work and, according to Stanley Heckadon, Panama's director general of the Institute of Natural and Renewable Resources, resulted in a massive migration from the country's cities to its forests. Here the Panamanians are clearing land to grow crops, and in some cases they are cutting in the national parks. Aside from the obvious effects of this activity on the nation's forests, the runoff from stripped land may be jeopardizing canal watersheds. We focused on Heckadon's concerns and, with them, the reporter's experience as he traveled through the Panamanian forests. Naturally, a balanced report would include interviews with officials from the Panama Canal Commission and the U.S. military. We did the interviews (with Canal officials—the military would not talk with us), but we did not put the officials on camera. Rather we incorporated their perspective—where appropriate—into the body of the program. We made a clear, conscious choice to focus on Heckadon's point of view.

It would have been inappropriate for the program to make a value judgment about whether the United States should or should not have invaded Panama, and we didn't. But we did direct audience attention to the issue of war's effect on the environment. By going to war, Americans make implicit choices, and in the next decade, if not in the next few months, it is worth remembering there is an environmental impact to war—and it isn't necessarily confined to the scene of the conflict. This is exactly the message that came through by focusing on Heckadon's experience in the forests.

ANOTHER PROBLEM WITH "balance" is that it is often artificial, a matter of giving equal airtime or newshole space to dissenting views of questionable merit. Global warming is perhaps the issue that has been most muddied by this practice. While there is rarely 100 percent certainty in science, an overwhelming majority of scientists believe that the theory of global warming is essentially correct, some would say irrefutable. Regardless, journalists have felt compelled to seek out contrary points of view, in some cases calling on experts of doubtful expertise and motive. The net result has been confusion among the American public. Who is right? With a "balanced" report the audience is left with more questions than answers. There are some in the environmental community, such as Lester Brown, head of the Worldwatch Institute, who believes we are wasting precious time creating this kind of confusion.

It may be more effective to report, as we do on "Network Earth," that global warming is a reality and then offer possible solutions. This is advocacy journalism as information broker. I do not have time to follow the stock market, so I hire a stockbroker. We broker the environment. Our audience does not have the time to follow all the developments in the global environment or to know about the dangers of pesticides or hazardous chemicals. Our job is to do the "buying," to provide interpretations and solutions. If we continually offer a balanced perspective, we haven't encouraged the audience to *do* anything.

Don't underestimate the extent to which the public wants this sort of guidance. There is incredible confusion and even terror among Americans concerning environmental issues; they are frightened their water is polluted, frightened their children may be poisoned by pesticides, frightened nuclear waste is at their doorsteps. A lot of these issues are making front pages because of fear. People don't understand how these environmental calamities happened in their communities, and they don't know how they can change them. At some point, balanced journalism simply does not give them answers; it gives them issues. It doesn't say, "Here's the problem and here's the solution." I am not suggesting that we create the news. But if you offer a path for change and concrete solutions, you take a step in countering apathy and promoting action.

Individual Americans show such fortitude in being able to turn things around that I think they are a tremendous, largely untapped source of power for environmental change. It is easier to inspire individuals to change their behavior than it is to get a corporation or the government to make modifications in its treatment of the environment. More important, it is public demand in this country, particularly at the local level, that has been among the most effective forces in getting corporations and public officials to take action on behalf of the environment.

StarKist is a prime example. Responding to public outrage over tuna-fishing practices that resulted in the deaths of thousands of dolphins each year, the company announced it would no longer buy tuna that were caught using drift nets or other methods known to kill dolphins. This was a dramatic victory for American consumers. They used the power of public opinion as well as a commercial boycott to effect change.

Elsewhere, throughout America, there are increasingly organized and dedicated groups that, through their concern for their communities and the quality of life there, are transforming environmental policy at the state and national levels. That is empowerment.

I WOULD NEVER assume that by being an advocate, a journalist is absolved from following the common guidelines of fairness. Your facts must be secure, and you must be ready to defend them. Neither does advocacy lessen the quality of the report or the journalist. Remember Edward R. Murrow's "Harvest of Shame"? But advocacy does mean that reporters take a more personal point of view. It may not be balanced, but it is quality journalism.

My greatest fear is that the advocacy position will be interpreted as P. J. O'Rourke characterized it in a recent article in *Rolling Stone* magazine, with bad guys and good guys, a world in which industry is some kind of amorphous creature out there destroying the environment. Advocacy does not mean Greenpeace is always right and the oil companies are always wrong. To begin with, *we*— the American public—are industry. I turn my lights on, I use my TV set, I drive my car, I use chemical products. Without "me" there is no industry. Without a sense of our circular responsibility in environmental degradation, advocacy won't work.

Therefore, the environment provides the news media with a very specific challenge. At what point does personal involvement cloud the objective telling of the issues? It's a hard line to draw these days. If as a journalist you spend your working life investigating environmental issues, you do at times go home with feelings of tremendous desperation. I have. When you see a town with an unusually high incidence of childhood leukemia and you hear people denying that it's a problem—that the toxic waste dump has nothing to do with it—it's an outrage. I don't know how people wake up in the morning in Los Angeles in August when there's a brown haze engulfing the city and not feel desperate and angry. It's a matter in many cases of simple greed, and not just of industry and government, but of individuals too. How do you turn some of this around? Perhaps by not simply reporting the facts, but telling the truth.

What's more, if you're going to focus on solutions, even in the mainstream news media, how are you going to do it and not take on some big targets? How are you going to cover the environment in any truly informative and socially responsible way without being perceived as an advocate, whether by your readers, your editor or your publisher? It's not surprising that environmental reporters, even at some of the nation's leading news organizations, have been asked to resign or move to another beat.

A single television program with an emphasis on advocacy is certainly not a panacea for our environmental problems. It can only be one voice among many. Yet now does seem to be the time for rethinking some of our journalistic canons. The "balanced" report, in some cases, may no longer be the most effective, or even the most informative. Indeed, it can be debilitating. Can we afford to wait for our audience to come to its own conclusions? I think not.

Jessica Tuchman Matthews, vice president of the World Resources Institute, told Bill Moyers in a 1989 interview that we have only 10 years "to turn things around," to head off trends in global environmental destruction. Those 10 years may come and go, and we may not be the victims of massive environmental disaster. But what if we are? What if "disaster" comes in a form that makes it hard to recognize: more cancers, fewer new medicines, escalating infant-mortality rates? Will we still want to "study the problem" some more? How long can we remain dispassionate?

I suggest taking out a journalistic insurance policy—information with a message, a message with a solution.

The 1985 sinking of the Greenpeace flagship Rainbow Warrior *by French government agents created a scandal that nearly unseated the Mitterand government.*
John Miller/Greenpeace

JIM DETJEN

The Traditionalist's Tools
(And a Fistful of New Ones)

———

In the summer of 1974, John Harris-Cronin, a dedicated environmental activist, and I paddled by canoe down the Fishkill Creek in Dutchess County, New York. As we floated by an ancient and foul-smelling factory owned by Tuck Industries, he pointed out 27 pipes from which oils, detergents, latex and other chemicals were oozing and dribbling into the creek, a tributary of the Hudson River.

The company, which made adhesive tape at the factory, was required by federal law to have permits from the U.S. Environmental Protection Agency for each of its discharge pipes. But a quick check of EPA records showed the company had filed for only two permits. The case against Tuck seemed airtight, black and white; they were clearly violating the federal Clean Water Act.

On the basis of evidence gathered by Harris-Cronin and other "river rats" who worked for the People's Pipewatch, a volunteer environmental group, the U.S. Attorney's office in the Southern District of New York charged the company with 24 violations of federal water pollution control statutes on June 5, 1975. The company at first pleaded not guilty. But then on Feb. 17, 1976, it admitted violating 12 counts of federal statutes and was fined $43,500.

Looking back at that episode a decade and a half ago, I am struck by how remarkably simple that case seems now—a classic good guy vs. bad guy, white hat vs. black hat encounter. Things are rarely so simple these days. Environmental controversies have become increasingly complex, and the gray areas have proliferated.

Fortunately, the caliber of journalists covering environmental issues has also improved. Fifteen years ago few environmental reporters had taken a single course in ecology. Today many have taken advanced courses in biology, chemistry, law and investigative reporting. Some have attended special seminars on environmental risks. Many have studied science and the environment in college and even in graduate school.

The number of journalists covering the environment has also mushroomed. While the exact number is hard to pin down, there is ample evidence that this is no longer just a tiny field. I am constantly amazed at the number of reporters who approach me after I speak at journalism conferences and say they have just been assigned to the environmental beat. Many of the newcomers work at television stations.

A recent survey by the Scientists' Institute for Public Information (SIPI), a nonprofit organization set up to improve science reporting, found that environmental coverage has increased by 72 percent in small newspapers during the past 2 years. The Center for Media and Public Affairs in Washington, D.C., reported in April 1990 that the number of environmental stories on the networks soared from 130 in 1987 to 453 in 1989. And *Executive Trend Watch*, a corporate newsletter, reported in 1989 that environmental coverage in newspapers and broadcast stations increased from a "steady 2 percent" nationwide in 1988 to 5.2 percent in the fall of 1989.

But is this coverage good enough? Is the quality of these stories consistently high? Are the media being aggressive enough in digging beneath the surface of stories such as the thinning of the ozone layer, the buildup of greenhouse gases in the atmosphere; the destruction of wildlife habitats; and the loss of topsoil, plants and animals?

The answer to each of these questions is unfortunately no. Environmental coverage at even some of the nation's best newspapers and broadcast stations is often timid, crisis-oriented and inconsistent. A top news organization may do an excellent job one day in reporting about an important environmental issue and then completely ignore an important development weeks later. Followups are few and far between.

Because of the concern by some activists and scientists that the media are not doing an adequate job, a few have called on the media to abandon their traditions of "objective" reporting. They say that the global environmental problems faced by society during the coming decades are so staggering, and will require such fundamental shifts in the way people live, that traditional approaches are no longer enough. They argue that journalists must become "advocates" if society is to prevent global warming, the continued decay of the life-protecting ozone shield, the continued loss of topsoil and forests, the degradation of the oceans, and the projected loss of 1 million species of plants and animals between now and the year 2020.

One of those who favors "advocacy journalism"—the abandonment of such traditional practices as "objectivity" and "balanced reporting"—is Barbara Pyle, environmental editor for Turner Broadcasting System. "We are running out of time," she says simply. Another is Mark Hertsgaard, a journalist who wrote in *Rolling Stone* on Nov. 16, 1989: "If our children, to say nothing of their children, are to inhabit a survivable world, some very fundamental changes have to be made. The success of this transformation will depend in no small measure on the news media."

Lester Brown of the Worldwatch Institute puts it this way: "We don't have time for the traditional approach to education—training new generations of teachers to train new generations of students—because we don't have generations, we have years. The communications industry is the only instrument that has the capacity to educate on the scale needed and in the time available."

While I agree that the media's coverage of environmental issues is often sorely lacking, I don't think advocacy journalism is the answer. I believe that advocacy journalism, if it means one-sided and unfair reporting, is misguided and in the long run counterproductive. If major newspapers, magazines and broadcast stations adopt an advocacy philosophy, the media will be treading on dangerous ground that could alienate readers and viewers and cause them to stop trusting the media. Journalists who have spent their careers establishing reputations for fairness and accuracy could suddenly find their credibility evaporating.

To be sure, objectivity is always hard to achieve. Diane Dumanoski, an environmental reporter for the *Boston Globe*, said at a conference in Minneapolis in May 1990 that there is no such thing as "objective" reporting. "From the very minute a reporter and editor decide to do a story, they have made a value judgment— a choice to do story A rather than story B." But, nonetheless, I believe it is vital for journalists to strive to be balanced and fair, reporting conflicting views even when they don't agree with them.

I believe that traditional approaches of the American media— such as investigative reporting, agenda setting and mass education—can do the job. It is important for the media to maintain their tradition of healthy skepticism, continuing to question government, corporations—and even environmental groups.

Readers have a right to expect that the media are performing their responsibilities fairly and without bias. Since the public does not know whether or not the media are, in fact, reporting the news objectively, they must trust the stewards of the institution to ensure that those who write and edit the news are not improperly influenced by financial, ideological or other personal considerations.

But while I favor traditional standards of objectivity and fairness, I believe significant changes are needed in the way the media cover the environment. Executives need to place greater emphasis on environmental reporting and provide journalists with greater resources to do their jobs. The media must become less provincial and more global in their coverage. Journalism educators should improve training in environmental and science journalism and offer midcareer fellowships specifically geared to environmental issues.

Some of the changes I'd like to see are these:

Expanded coverage of environmental topics and greater resources made available to journalists covering the environmental beat. Tom Winship, former editor of the *Boston Globe,* said in a speech he gave on the 25th anniversary of *Columbia Journalism Review* in New York City that the press "should assign the same news values and resources to the global environmental beat as it does to the nuclear arms race and to national politics." I believe that's true. Most assigning editors in the media come from traditional beats such as police news, local government and politics. They, in turn, usually funnel much of their news organization's resources into these subject areas. They may assign several reporters to cover politics on the local, state and national levels. But how many do they assign to the environmental beat? Usually just one.

Newspapers and broadcast stations should assign more reporters to this vital beat. The smaller paper without an environmental reporter should hire one. The larger paper that has one person covering it should consider establishing a special team reporting about local, national and international developments. Why not an "E team" at larger papers, that regularly investigates air and water pollution, nuclear power plants and chemical companies? Poll after poll shows that the public cares deeply about the environment and wants to read and see more about this subject. Increased environmental reporting would attract more viewers and readers, particularly among the young who are not the avid consumers of newspapers that their parents and grandparents were.

Expanded coverage will also serve the best interests of the

community because increased reporting will almost inevitably lead to greater pressure on politicians to clean up Superfund sites, reduce air and water pollution, and preserve forests and wetlands—efforts that benefit everybody.

The media should also invest more money in training environmental journalists. Environmental reporting is complicated, requiring knowledge about biology, chemistry, geology, meteorology, statistics, public health, law, government and other subjects. More journalism schools should set up degree programs in environmental reporting. Midcareer fellowships should be established to teach journalists about new developments in toxicology, radiation, environmental risk and other complicated areas.

Greater consistency and less crisis-oriented coverage. One of the biggest continuing problems in environmental reporting is the lack of consistency. During the mid-1970s, when interest in the environment ran high, many papers started full-time environmental beats. But as the public's interest seemed to wane in the early 1980s, many papers eliminated or cut back on this beat. As a result, some of the important environmental stories of the 1980s—such as regulatory cutbacks at the U.S. Environmental Protection Agency—received minimal coverage.

It is not unusual to find journalists who have spent several decades reporting about national politics or baseball or finance. But it is rare to find reporters who have written about the environment uninterrupted for the past 2 decades. Casey Bukro, the environmental writer for the *Chicago Tribune*, is one of the few exceptions. We need more veteran environmental reporters who have the perspective that comes only with experience.

The media spend too much time chasing after fires, reacting to daily crises, and not enough time examining long-term problems. Much has been written about how the national media largely missed such important post–World War II stories as the migration of millions of black Americans from rural to urban areas. A major reason for this is that blacks did not hold press conferences to announce they were moving.

The same situation exists with environmental stories, which by their very nature are often gradual, long-term trends. Community activists protesting the siting of hazardous waste incinerators or nuclear dumps get a lot of coverage because the activists are vocal and have media savvy. But much less coverage is given to long-term environmental stories such as the decline in migratory song-birds in the United States, the erosion of topsoil, and the increasing haziness that blankets large areas of the nation during the summer months.

Editors need to give their reporters greater latitude in reporting about long-term environmental stories. They should give reporters the time to step back from the pressures of daily journalism and the chance to discern the broader trends. It's a lot easier to sell editors on dramatic stories such as the Exxon *Valdez* oil spill than it is to convince them to report about an academic theory predicting what might happen 30 years in the future. But scientific theories such as global warming may in the long run turn out to be vastly more important.

One way to improve consistency might be to develop a measure—let's call it the Environmental Index, or EI—that the media could report each day. Newspapers and broadcast stations are fascinated with daily numbers, such as the Dow Jones industrial average, the price of gold and the scores of last night's baseball games.

Realizing the media's predilections for such numbers, I believe that scientists should develop similar measures for the environment. An environmental index for each metropolitan area, and perhaps a national and global index as well, could help keep track of the state of the environment and serve as a constant reminder of what remains to be done.

I'm not an ecologist or statistician, so I don't know how difficult it would be to develop such an index. I'd prefer to see a daily index, based upon information gathered by the EPA, the National Aeronautics and Space Administration and other federal agencies using satellites, air and water pollution monitors and other sensing devices. If that's impossible, perhaps an index could be developed that reports weekly or monthly findings, similar to the nation's economic indicators.

The EI might include information about a variety of environmental subjects: ozone, sulfur dioxide, nitrogen dioxide, lead and other air pollutants; levels of ultraviolet radiation reaching the Earth's surface; carbon dioxide concentrations in the air; counts of biological oxygen demand, fecal coliform counts and other measures of the health of the nation's waterways; topsoil erosion; and the populations of wildlife.

In a limited way some of this is already being done. Many air pollution control districts provide the media with daily readings of ozone and other air pollutants. In Australia, where chlorofluorocarbons have dramatically thinned the ozone layer, some television stations report the daily ultraviolet readings so that people know whether to go to the beach or not. The Worldwatch Institute publishes surveys on global environmental matters in its annual "State of the World" reports.

If such an environmental index were available, I believe that many journalists would report its findings regularly, much as they keep track of the weather today. The impact could be substantial. The EI could focus greater public attention on the environment, much as Walter Cronkite kept viewers focused on the hostage crisis in Iran by commenting at the end of his evening broadcasts a decade ago, "And that's the way it is, the 167th day of American hostages in captivity."

Increased coverage of global environmental problems. In the early 1970s, journalists focused primarily on local problems such as oil spills contaminating harbors or smoke fouling the air downwind from a coal-fired power plant. The concept that the actions of one nation could damage the ecosystems of another seemed remote and largely unproven.

Today all that has changed. Scientists understand now how interconnected the world's environment is. CFCs sprayed from an aerosol can in Canada have eroded the ozone layer above Antarctica. Toxic wastes produced in Illinois and transported across the Atlantic Ocean have polluted the drinking water of Africans. Forests cut down in Latin America have destroyed the wintering habitats of North American songbirds.

As scientific understanding has increased, technology has advanced. With lightweight video cameras, communications satellites and jet planes, journalists can fly to a distant site in the morning and broadcast a report home live for the 6 o'clock news. While some newspapers and television stations have begun to send their reporters around the globe to work on environmental stories, the numbers are still limited. But if more media executives supported journalism projects on global environmental issues, awareness about the urgency of these problems would increase.

"Sixty to 70 percent of acid rain in Canada comes from the United States. All the Canadians know it, but the Americans tend to forget it," said Alfred Thorwarth, the on-air host and writer for the television program "Globus," a half-hour West German television program that focuses on environmental themes. "The Japanese have some of the best national parks in the world, and they are cutting down every single tree in the Pacific basin. There's still the attitude of, 'If it's not my country, I couldn't care less.' We as journalists have a tremendous responsibility to change this sentiment."

Expanded investigative and project reporting. Media in the United States have a long tradition of muckraking about political corruption, the improper use of police power, safety hazards and government bureaucracies run amok. The tradition is much more modest when it comes to investigating environmental problems. But it is growing.

In 1990 two of the Pulitzer Prizes were awarded to investigative projects on the environment—a series by the *Seattle Times* won the national award for reporting about oil tanker hazards, and stories about carcinogens in the drinking water of Washington, N.C., won the public service award for the *Washington* (N.C.) *Daily News*. In addition, five of Sigma Delta Chi's Distinguished Service Awards went to environmental reporting, and environmental exposes accounted for two of the six newspaper awards given by Investigative Reporters & Editors for that same year.

In his 1986 speech at Columbia, Winship advocated greater efforts to investigate environmental problems. "For years we've had gumshoe teams investigating corruption," he said. "Now, it's time the big healthy papers create an integrated team of specialists to monitor regularly the mauling of our surroundings."

Among the promising new tools being used by investigative journalists are computer data bases. Scott Thurm, the former environmental reporter for the *Louisville Courier-Journal* (and now with the *San Jose Mercury News*), utilized an EPA data base to write several stories describing the types and quantities of toxic chemicals being stored at Kentucky chemical plants. Other journalists have used computers to analyze pollutants being discharged into the nation's air and water.

One development expected to spur investigative environmental reporting is the formation in February 1990 of the Society of Environmental Journalists, the nation's first organization dedicated to environmental reporting. Through a newsletter, workshops and conferences, the society hopes to train reporters to use the federal Freedom of Information Act, computer data bases and other investigative tools to probe these kinds of stories.

Increased reporting about alternatives. Another problem in the media's coverage is their failure to write about alternatives. In addition to investigating environmental problems, journalists need to show readers that there are solutions to these problems. By reporting about efforts to conserve energy, recycle wastes and save water in other communities, journalists can perform a valuable educational role.

I make an effort to tell readers about concrete steps they can take to solve environmental problems. For example, in a *Philadelphia Inquirer* magazine article on global climate change published on April 29, 1990, I listed steps readers could take to protect the ozone layer. These measures included buying fire extinguishers that don't contain halons, ozone-depleting chemicals, and making sure their cars' air conditioners are not leaking CFCs.

Is this advocacy? I don't think so. I am simply presenting readers, who have a hunger to take concrete actions, with practical tips or "news they can use." When I have written stories about scientific research on lightning or tornadoes, I have also included practical tips on how to avoid injury from these severe weather phenomena. Some might call this advocacy; I prefer to think of it as public service.

I think it is irresponsible for journalists to report only about gloomy environmental developments. I think it is important to show readers that there are practical steps they can take to conserve water and energy, prevent environmental degradation and save themselves money.

NONE OF THESE five suggestions involve the abandonment of traditional standards of fairness and objectivity. Yet, if implemented, they could go a long way toward improving the public's knowledge about environmental problems and providing information on how to correct them. Traditional aggressive reporting, not advocacy, is the key.

Or, as Bob Engelman, an environmental reporter for the Scripps Howard News Service, puts it: "The challenge is the same as for any [issue] that is complex, important and likely to remain in the public eye: Learn the issue. Maintain skepticism. Seek out all viewpoints. Ask probing questions. And report the story as accurately and fairly as you can."

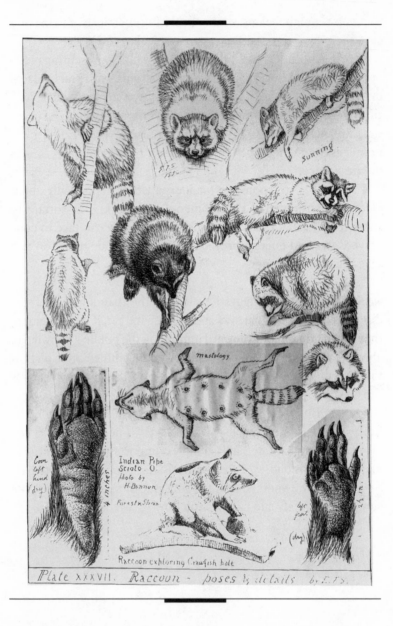

Raccoons by Ernest Thompson Seton
Philmont Scout Ranch, New Mexico

CRAIG L. LAMAY

Heat and Light: The Advocacy-Objectivity Debate

Allan Hammond, the Yale nuclear physicist and former editor of *Science ('84, '85, '86, etc.)* magazine, was once asked about the difference between scientific writing and journalism. He replied that, in his experience, scientists were constantly trying to turn magazine articles into scholarly ones for fear that their peers would look disdainfully at the inevitable generalization and simplification that occur in journalism. The journalist, Hammond said, is interested in context, while the scientist is interested in content: The journalist's goal is to inform, the scientist's to teach.

To many people, including journalists and teachers, this distinction is semantic hair-splitting. But as an editor who enlists writers from the ranks of scholars, journalists and policy-makers, I think I know what Hammond was talking about, and I would extend his distinction to hold generally between academic argument and journalistic writing. The magazine I edit is a journal of informed opinion; and unless they're practiced, neither academics nor journalists are comfortable with editorializing on their subject matter—not in print anyway, and not without encouragement. Both are more at ease with the idea of building an argument than with entering into one.

Academics, of course, do enter the arguments they build—but cautiously, analyzing every turn with the methodological tools of their discipline, aware that their foray into interpretation is the last critical leg of a journey in which they risk their scholarly reputations. Journalists are another matter. Some are so averse to debating the merits of the arguments they've constructed that you may as well ask them to amputate one of their own limbs. What they do instead is heap more facts on the fire. They guard their professional reputations by studiously avoiding argument. Here lies the distinction I think Hammond was suggesting: Academics can be said to teach because they sift through their arguments for the spurious and the irrelevant in order to build a *better* argument; journalists, dedicated to enhancing what Walter Lippmann called "the machinery of record," simply build a *bigger* argument—they inform.

This difference, I believe, defines the point of contention between environmental reporters and editors who argue about "objectivity" and "advocacy." The objectivity-advocacy debate has become a fixture at environmental reporting conferences, a time-consuming and unfortunate distraction that has gained currency largely because a few well-known journalists have been induced to make revival-tent declarations of their environmental mettle. At the center of the argument are reporters and editors who say it is their duty to "empower" their readers, to give them answers to problems; these journalists are opposed by those who ask (confidently and rhetorically) if the environment is really any more important or deserving than, say, education or foreign policy.

Anyone who has ever sat in on one of these discussions soon senses that, except perhaps at the extremes, the polemics don't quite join at a clear point of contention. They don't. Those who promote advocacy conjure up images of human and environmental horrors and insist, rightly I think, that such things are ill-served by the journalistic habit of building a bigger argument. They may want more environmental coverage and more visible coverage (what reporter in what specialty wouldn't?), but few seriously expect that to happen or, if it does, to be sustained. Most reporters are career skeptics, and all are aware of the ephemeral nature of news and the demands of the news marketplace.

But reporters who promote environmental advocacy do not assail journalism in the usual way—charging that a lack of "balance" has resulted in media "bias"—because they do not believe balance is the problem. They want a better argument, not a bigger one, and in trying to fashion that argument they have broken from the accepted lexicon of journalistic debate and asked for something truly radical: the freedom to arbitrate the arguments they construct. It's not articulated as such, but the advocates feel it is their duty to teach, and the objectivists are constitutionally unable to join this argument except by dismissing it. This dismissal is why these discussions are so heated and unproductive, like gears grinding away, unable to engage one another.

But despite the heat there may be some light, for the debate has implications for journalism generally. Environmental reporters are simply the most vocal members of a larger group of reporters from many other beats, all of whom are beginning to feel that the information marketplace as it now operates doesn't work very well. When even reporters who get such career-making assignments as the White House or presidential campaigns say, as they have for several years now, that they are simply the objects of someone else's information management strategy, it's time to look again at the machinery of record. At least one other author in this book argues that the nation's information system has been so corrupted that the machinery of record can't hope to produce anything of value from it, and I suspect many journalists from many beats agree with her, though they may be reluctant to say so. To admit it, after all, would be to lose one's professional raison d'être.

News executives, however, know very well that the information marketplace isn't as profitable as it once was. Since the early 1980s the evening news programs on ABC, CBS and NBC have lost millions of viewers, and while many of those viewers now watch cable channels like CNN and HBO, or perhaps public television, the overall television news audience is shrinking and has been for nearly a decade. Newspapers are also worried about their future. In 1965, a Gallup poll found that 87 percent of Americans under 35 said they had "read a paper yesterday," but only 30 percent said that in a June 1990 *Times Mirror* survey. According to *Times Mirror*, young Americans aged 18 to 30 know less and care less about news and public affairs than any other generation of the past 50 years. Together, newspaper and broadcast news audiences in America have deteriorated to the point where they consist of fewer individuals than the number that vote in most elections.

What has happened to conversation about public issues? Writing in the spring 1990 issue of the *Gannett Center Journal*, historian Christopher Lasch wrote that "What democracy requires is public debate, not information":

> Of course [democracy] needs information too, but the kind of information it needs can be generated only by vigorous popular debate. We do not know what we need to know until we ask the right questions, and we can identify the right questions only by subjecting our own ideas about the world to the test of public controversy. Information, usually seen as the precondition of debate, is better understood as its by-product. When we get into arguments that focus and fully engage our attention, we become avid seekers of relevant information. Otherwise, we take in information passively— if we take it in at all.
>
> From these considerations it follows that the job of the press is to encourage debate, not to supply the public with information. But as things now stand the press generates information in abundance, and nobody pays any attention. . . . Since the public no longer participates in debates on national issues, it has no reason to be better informed. When debate becomes a lost art, information makes no impression.

Lasch attributes a portion of the lack of public debate to the press itself, to its 20th-century embrace of the professional objectivity articulated by Walter Lippmann. Another portion he attributes to the information peddling of the public relations and advertising industries, which simply filled the void the press created when it became objective and "responsible." In the information society, Lasch concludes, advertising and publicity substitute for open debate, and the predictable result is a society of competing private interests in which there is no articulated notion of the public good, or even of the "public." For many journalists, and certainly for politicians, the public is a necessary fiction that gives structure to a fractious and endlessly segmented audience, a marketplace of desires. Media organizations and political parties are endlessly reinventing their "products" to meet the demands of this marketplace—usually with the effect of fragmenting it further—but they give little thought to the values necessary to create and sustain a "public."

In the 1980s the Reagan administration cynically exploited this marketplace mentality to help destroy the idea of a public—of a common culture—as an impediment to self-interest. The clearest indication of its success is that the language of a common culture is now anathema to the left as well as to the right, a phenomenon that the media have done a lot to enlarge but little to illuminate.

They have done so at their own peril, for without something of a common culture, however it is defined, the importance of mass media declines. In his book *Three Blind Mice*, journalist Ken Auletta says that the decline of the Big Three networks' dominion over broadcasting is a blow to the concept of shared national experience, but it could be just the other way around—that when it was no longer legitimate to think of a shared national experience, the networks lost some of their legitimacy too.

In such a political climate, journalists who argue for advocacy may be ahead of their peers in recognizing that their profession needs to look critically at where it is going. If journalists don't, they may find themselves toiling as monks in a new age of information elites. What they want is within their grasp. They know full well there is a debate already going on, one that includes, say, Exxon, the EPA, the Sierra Club and everyone in between—everyone but them. For reasons of tradition and institutional caution, reporters are expected to monitor debates, to record, clarify and simplify them, but not to enter them, judge them or, even, ignore them; this implies as well that they cannot transform them by deciding what does or does not merit coverage. Their hands tied, journalists in many fields say they often feel like glorified publicity agents. Not coincidentally, communications scholar Scott Cutlip has estimated that some 40 percent of all news stories originate not from reportorial initiative, but as press releases from organizations and individuals who want to get their message into the media system. From the standpoint of those carrying on the debate, that's the idea. In their hands, "objectivity" is little more than a nightstick that they have wrested from its owners and used to subdue them—or at the very least to control the flow of information in the system.

Many if not most environmental advocates premise their arguments, of course, on their belief that journalistic objectivity is nonsense. Other essays in this book make the case for that point of view, and one need only watch 5 minutes of the evening news or skim the front page of the newspaper to see the substance of their claim. When the news media uncritically and regularly report the frenzy of the nation's consumption as a measure of its economic and social health, for example, the practice speaks volumes about the prevailing values that shape the news. They are objective only in the sense that they represent society's dominant values, largely agreed upon by government, commerce and other institutions—including the media. Objectivity as consensus, in fact, is essentially the information paradigm that Walter Lippmann helped to establish in the 1920s. If anything, it is a kind of Orwellian notion in which consensus belongs to those with the power to make it, where a thing or a process or a person is whatever the powers that be decide to call it.

But while the news system may operate according to a limited set of values, it is neither ideologically predisposed nor closed. It may force divergent points of view to accommodate their messages to the media "market," but the messages can get into the system. Robert MacNeil, who has spent a good part of his journalistic career working against the centripetal pull of the media marketplace, has said that journalists can set their own agenda only if they choose to work out of the mainstream. And indeed, in environmental coverage, as in many other areas, the best sources of information are specialized media—newsletters, trade publications, electronic bulletin boards and other services—that do not serve a mass audience. But of course many environmentalists and the journalists who cover their work feel that environmental issues need to reach a broader audience and that the mass purveyors of objectivity are, unchallenged, doing the public more harm than good. So they work to tailor their messages to the conflict-and-personality needs of the system. The question is, with what result.

When the Natural Resources Defense Council (NRDC) wanted to publicize the dangers of Alar, for instance, it enlisted actress Meryl Streep as its spokesperson. The news media dutifully gave her equal treatment with scientists, and the public was asked to choose between the two. Aside from the question of Alar's dangers, what was the NRDC's campaign intended to do? Was such an obvious publicity venture the only way the NRDC could get news attention for its views? In a market-driven information system in which publicity is the highest good, where personality and conflict are the sources of news, intentions don't matter and no one asked. But the questions remain: Was this good reporting founded on good evidence? Was it reporting in the public interest? And if the NRDC's strategy was ill-advised, as many think it was, what is the alternative?

To teach, for one. This is essentially what many environmental reporters mean when they argue for being advocates: they want to enter the debate, to debunk spurious claims, to arbitrate analysis, to guide reasoning. They want to make information actionable.

But there are problems with this ambition. The first is that there is nothing in most journalists' training or background that imparts to them the cultural authority to speak as teachers. Most are trained as generalists whose specialties, if any, are gained through seat-of-the-pants experience rather than formal education, and while there are always those who can step beyond generalization—Bill Moyers comes to mind—the exceptions are often ridiculed for what other journalists see as their presumptuousness. There is likely a tinge of envy at work here, though, for almost every journalist, no matter how circumspect, aspires to instruct. During the Persian Gulf War, I heard a veteran war correspondent say as much, plainly and publicly, lashing out at a respected Middle East scholar for his "pomposity" and declaring that most experienced reporters—of which there were very few in the Persian Gulf—knew as much or more about the military aspects of the affair than most academics. I have heard other reporters from other beats say essentially the same thing—privately—and I suspect that their feeling devalued contributes to the tension that almost always exists between journalists and scholars.

Other problems with teaching arise from the fact that the environment is such a broad topic, with roots in the humanities—philosophy, history, economics—and the natural sciences. Who can presume to teach? Few reporters can; those with long professional experience and substantial scholarly backgrounds are the first to recognize the scope of the job, and most of them rely on experts in narrower specialities for their information.

Teaching is also a highly politicized endeavor these days, having succumbed to the same common culture controversies that have undermined public life generally, and much of its intellectual product, particularly in the natural sciences, is proprietary, funded and owned by private corporations. (Take note the next time you hear scientists debating environmental issues on news programs or see them cited in print articles; reporters almost never ask them about their funding sources.) A field like environmentalism, with roots throughout the disciplines, is necessarily a tangle of competing ideological and commercial interests. Arguably the most promising academic field for environmental thinking is economics, where politics are less an issue than are methods and values that pretend to the precision of science.

So what to do: to presume to teach or to retreat to inform?

Morally there is no choice: better to teach and navigate through oceans of information than to inform and simply drown in them. Sidney Hook once wrote that "moral responsibility in history consists in being aware of the relevant ifs and might-be's in the present, and choosing between alternatives in the light of predictable consequences. . . . If there is any ethical imperative valid for all historical periods it is awareness and action."

To teach, obviously, will require a system of values apart from traditional news values. The "ethical imperative" that Hook speaks of suggests moral values that are informed through intellectual effort, that can give information structure and make it actionable. In those terms teaching is exactly what the mass media *do not* do; they call themselves teachers, because otherwise they are little more than production houses, but teaching is really incidental to their purpose. Indeed, there is perhaps no more *amoral* institution than the mass media.

I think it possible that the advocacy movement in environmental reporting may indicate, rather than reporters' inability to separate themselves from their work—of which they are often accused—their wish to distance themselves from it even further. While obviously this is not true in every case, for many the goal is to think about and develop a new system of values that they can put between themselves and the information they gather so that they can filter it, structure it—and make it possible for the public to act on it.

This is an enormous task, and will require bright minds and courageous leadership. But it is not impossible, and it is not so different from what reporters on other beats want—virtually all reporters know much more than they say publicly, and almost all of them, when asked, will privately tell you the "real story," whether it concerns a town councilman, a Superfund site or a military mission.

It is time to rethink Lippmann's dictum that we should concentrate on improving the "machinery of record" rather than on educating the public; it is not enough to distill conditions into their constituent facts. As anyone who reads a daily paper knows, such pseudoscientific reductionism is impossible; disagreements occur among the most expert and systematic researchers. To present such disagreements as the end of inquiry is disingenuous, and many reporters know it. As John Dewey, Lippmann's contemporary, pointed out, systematic inquiry is only the beginning of knowledge, not its final form. To behave otherwise is to shackle willingly the hope of knowledge, to reduce all questions to political issues in which "objectivity" belongs to anyone with the proper credentials and where all "results" are equal until they have been tested either by the whims of economic markets or in the court of public desires.

Environmental reporters who promote advocacy simply want to push their inquiry further. They argue, in a sense, that they should say what they know and that their doing so will have a public benefit. Importantly, I think, they mean public writ large, not as an audience but as an actor. It is not their intention to divide the world into an "us and them" story, nor do they uniformly insist that environmental values should always be preeminent. Some, to be fair, do; but many disavow such views, which represent the kind of reductionist thinking, so common to news coverage generally and environmental coverage specifically, that they are trying to *avoid*.

My own hope is that the advocacy movement among environmental journalists will eventually have a creative impact on other news beats in the mass media. It has once before, in the 1960s. The economic transformation of the television news and entertainment marketplace that began in the 1980s is, for better or worse, still under way. Among other things, that marketplace is becoming truly global, a process that should give further impetus to thinking about the values that guide our media system. Newspapers have a longer history of virtually reinventing themselves to meet new demands, in major ways and small ones, and always when their critics thought they had expended the last of their energy and imagination—or lost their way altogether.

And the media will change again.

Intuitively, I suspect, many environmental reporters know this, and they know that change will bring them new opportunities to test their ideas, to reinvent the public as well as the news product. Drawing as it does from so many disciplines, their field offers journalists a unique opportunity to rethink many of their most debilitating professional practices, an exercise that could benefit all reporting.

Washington, N.C., residents taking water from a U.S. Marine tank after state officials declared the town's water unfit for drinking, cooking or bathing: September 1989
*Ric Carter/*Washington Daily News, *reprinted with permission*

WILLIAM J. COUGHLIN

Think Locally, Act Locally

On any newspaper, doing tough stories requires commitment and sometimes courage. But doing them on a small newspaper brings some special challenges. You write a story that criticizes a city council member, then eat lunch in her restaurant. You reveal something the mayor doesn't want revealed, then borrow money from his bank. You report mismanagement at the jail, though the sheriff's daughter is your son's classmate.

—*Charlotte Observer*
April 16, 1990

That editorial in the *Observer* was commenting on a series of stories in the *Washington* (N.C.) *Daily News* in fall 1989 that led to the cleaning up of cancer-causing chemicals that had contaminated Washington's water supply for more than 8 years. The series won for our 10,700-circulation family-owned afternoon daily the 1990 Pulitzer gold medal for distinguished public service.

But as the *Observer* was aware, a special problem exists for any small-town newspaper tackling a ticklish local issue. You can't always turn off your computer at night and forget the story when you go home. Someone you are assailing may literally live next door to you, and he is almost certain to have as many friends on the block as you do.

You are also likely to find on an environmental story that you are attacking one of your community's major economic interests, whether industrial or agricultural, present or potential. What that means becomes clear when your readers point out the hundreds of jobs you are putting at risk or the number of tourists or potential residents you are driving away. Those economic interests also may be among your largest advertisers, and don't think for a minute your publisher doesn't ruminate about that fact.

All of those things may be equally true in a large city, but the numbers just don't have the impact they do in a small town. When you're talking about a community of less than 10,000, a few dozen jobs or a few dozen more taxpayers can make a difference. The people you meet every day on Main Street know that just as well as they know who you are. You may think you are a hard-nosed editor in the proudest of journalistic traditions; they may think you're a jerk.

If you live in a largely one-party community in the South, the mayor you are attacking probably has been a friend and close political ally of your owner for many years. Or perhaps you want to go on an anti-smoking crusade in a major tobacco-growing area where the farmers and the advertisers who do business with them think your product is hazardous to *their* health.

Now look around your newsroom. Of the seven people on your news staff, just four are reporters and one of those is a full-time sportswriter. With the other three, how are you going to cover in a four-county area all the town and county council meetings, the obits, the church suppers, the Rotary and Ruritan meetings, the school board sessions, the cops, the courts, local industry, the new highway, the tulip festival, the election, the high school graduations and still go after that red-hot environmental story?

If you figure out how to do it and you do it well, are you prepared for the resolution before the city council honoring your newspaper to be voted down, 3–2? It happened to us.

WHEN IT COMES to environmental problems, Washington, a historic colonial community on the Pamlico River and the seat of Beaufort County, is not all that unusual as small towns go. It has a phosphate mining operation nearby that was fined $1 million by the state in 1989 for polluting the air, and a paper mill trying to clean up its act by eliminating the deadly dioxin it pumps into a nearby waterway. A much larger community upstream dumps so much untreated sewage into the river that state environmental officials say half of what flows under the bridge and past our town at low water is a mix of treated and untreated sewage.

Commercial fishermen in the Pamlico and Albemarle sounds are worried because the crabs, oysters, shrimp and finfish are disappearing as a result of water pollution and unregulated trawling practices. Not all of the problems are due to sewage or industrial pollution; chemicals used on nearby farms run off into the river when it rains, as do contaminants from the parking lots of riverside housing developments.

But Washington is not all environmental problems. It was one of the earliest colonial ports, and the churchyards in its historic district are sprinkled with graves of the first settlers. A pleasant park and promenade border the river. Its tree-lined waterfront attracts boats traveling the nearby intracoastal waterway and gives access to Pamlico Sound, one of the finest sailing areas on the East Coast. Washington also is a gateway to some of the largest and most isolated wildlife sanctuaries in the coastal wetlands, including the new 93,000-acre Alligator River refuge funded by the Mellon Foundation. Daily ferries from Swan Quarter in the next county link us to Ocracoke on the Outer Banks. In short, this is a town and a region of historic importance and economic promise.

In 1981, the superintendent of water resources in Washington, Jerry Cutler, worried by state and federal reports he had been reading on contamination of municipal water supplies, sent samples of the city's water off for testing. A private firm reported that the water contained more than nine times the maximum amount of cancer-causing trihalomethanes regarded as safe by the U.S. Environmental Protection Agency (EPA).

This was the first record of carcinogens in city water, and Cutler passed the information along in a memo to the city director of public works. At least three mayors, three city managers, the city council and state and EPA officials were to be informed of the problem before the public became aware of it, 8 years after that first lab report.

There were no federal or state regulations forcing the city to act on the information, and it did not. What appeared to be more pressing problems, including a sewage treatment plant that did not meet federal requirements, and cost overruns in construction of a new electrical substation, occupied the attention of city officials. They remained preoccupied when the local Coca-Cola bottling plant warned that Washington's water supply was contaminated. Nothing was said publicly, and Coca-Cola began carbon filtering of its own supply.

Congress had made provision for financial aid to communities to upgrade their sewage treatment facilities, but the Safe Drinking Water Act of 1979 did not offer any federal assistance for water plants. Furthermore, communities of less than 10,000 were not even required to test their water. That changed in January 1989 when towns of between 3,000 and 10,000 population were required by the EPA to begin quarterly testing. Washington, with a population of more than 9,000, fell into that bracket, although it still was not required to publicize the results. As the mayor later was to complain, "If we were under 3,000, they wouldn't be bothering us."

A notice on the back of utility bills in summer 1989 that the city was testing its water caught the attention of the *Daily News*. Reporter Betty Gray got a copy of the test results and began checking the list of 42 chemicals in the water with private and state toxicologists. She was told that one of the chemicals was carcinogenic and that it was in the water far in excess of the level the EPA considered safe. The local office of the state environmental agency, she learned, had been using city water as a control in checking other water throughout the region. Then she found the zinger:

memos showing that agency employees had been advised by their headquarters in Raleigh to drink bottled water because Washington water was contaminated. None of this had been made public. Because of the federal exemption for cities under 10,000 population, some 56,000 water treatment plants serving millions of people across the United States were not protected by the EPA. Of those plants, 104 were in North Carolina.

When she felt she had enough information to confront city officials, Gray met with city manager Bruce Radford in his office. Radford broke off their talk but promised to see her again that morning, then appeared in the newspaper office with a legal notice he wanted to run that day in the classified section. Told those pages already were on the press, he asked that it be run as a display ad. It also was too late for that. He left the notice on the editor's desk and returned to his office for the interview with Gray. The notice admitted publicly for the first time the presence of "levels of certain chemicals which exceeded EPA's recommendations" in the city's water supply. The front page was made over so we could lead with Gray's interview when she returned. That afternoon, reporter Mike Voss was assigned to join her on the story.

The next day city officials struggled to get their stories straight. In an interview with Voss, Radford said the city had known for 3 years of excessive levels of the cancer-causing chemical in its water supply, although it had not made the information public. But Mayor J. Stancil Lilley said he and the city council had known of the problem for "only a week or a little more," and then Radford appeared on television drinking a glass of city water to refute what a TV newscast called unconfirmed reports of contamination. That was before the city released test results showing levels of carcinogens in the water that averaged nine times the EPA's safety limit at 10 locations around town, including a drinking fountain on the second floor of city hall.

At the beginning of the following week, working from a telephone tip, Gray discovered that Coca-Cola had advised the city of water contamination even earlier than Radford had admitted.

The *Daily News* then obtained and published a state report that a second carcinogenic chemical had been found in city water that was combining with the first to produce such deadly concentrations that the water plant might have to be shut down. Mayor Lilley appeared at the newspaper's office to complain to the owner that its coverage was throwing the town into "turmoil," but his protests were too late. The state summoned city officials to a meeting to inform them that the combination of chemicals in the water was so dangerous that the cancer risk was an astounding 1-in-250, not the 1-in-10,000 that city officials had been citing or the 1-in-1,000,000 that the EPA and the state considered safe. State officials told residents of Washington not to drink the water; not to clean or cook fruits or vegetables in it; not to drink tea, coffee or lemonade made with it; not to take showers in it; and not to inhale the steam if they boiled it. City schools turned off their drinking fountains and switched to paper plates and plastic knives and forks. The hospital announced that if other measures failed, it would sink its own well.

On Saturday, September 23, just 10 days after the *Daily News* broke a story that city officials had been covering up for 8 years, a U.S. Marine Corps convoy rolled across the Pamlico River bridge into Washington and set up water wagons throughout the city to provide residents with clean drinking water. We looked like a Third World hamlet.

The rest was cleanup. The mayor and a majority of the city council were swept out of office in a municipal election in early October. The city manager later was asked to resign. The new administration took emergency measures, costing $14,000 a month, that enabled the state to lift its drinking water ban the day before Thanksgiving—just 2 months after it was imposed.

After similar contaminants were found in the drinking water of 13 other small towns, the state health commission changed its regulations to afford protection to all communities in North Carolina, whatever their size. The EPA is considering similar changes that would go into effect next year. State health director Dr. Ronald H. Levine, in interviews with the *Charlotte Observer* and Raleigh *News & Observer*, said the regulatory changes brought about by the *Daily News* could save hundreds of lives.

SO WHAT SIGNIFICANCE does all this have for other small newspapers trying to do environmental stories? What did we learn, and what will we do differently next time? My own view is that an environmental story is no different, no more or less important, than any other story a newspaper covers that affects its readers. A corrupt judge, a high infant-mortality rate, a tornado—aren't they all part of a town's environment? Defining the limits of an environmental story is about as easy as distinguishing between so-called investigative reporting and plain old-fashioned, ask-the-awkward-questions reporting that has always been practiced on good newspapers.

So, yes, we continue to do environmental stories to the limit of our resources, just as we do the other difficult stories. Currently, for instance, Weyerhauser, the tree-chopping and tree-growing company, has proposed a planned community of more than 850 homes on the south bank of our river, complete with golf course and marina. Environmentalists are fighting it in court because of possible water pollution. They expect us to be on their side because, after all, we have demonstrated our environmental mettle. I am more concerned with whether we are being fair to the other side. The addition of 850 middle- and upper-income families and a $250 million tax base to our community would seem to be a plus. And most assuredly many of them would become subscribers to the *Daily News*, meaning a not insignificant jump in circulation for us. Thus the issue of how we cover this story becomes somewhat more complex. Which brings me to some ground rules.

First, leave the crusading to the Knights Templar. You'll be better off if you keep in mind that your job is reporting, not crusading. As *Newsweek* said recently in regard to the S&L crisis, "In the end, voters have their own responsibility. The press can lead the horse to water. The horse has to decide whether to drink." It's easy to get

emotionally involved in environmental stories. The good guys and
the bad guys can be spotted quickly, right? Let me ask you a few
questions about our water scandal. Were the real villains the local
and state officials who did nothing about the contaminated water
because no regulations required them to do so? Or was it Congress when it passed a Safe Drinking Water Act that exempted
communities of our size because of the additional federal funds
that would have been required to protect them? Is it more important now for our community to increase taxes to eliminate carcinogens in drinking water that may give 40 people cancer over the
next 70 years, or to do something about the fact that this region
has the highest infant-mortality rate in the state with the nation's
worst infant-mortality rate? Are the two related? They may be, but
the newspaper can only inform. The public must decide.

Second, do your technical homework and translate it for your readers.
Carcinogenic trihalomethanes were being produced when organic material from Tranters Creek, a tributary of the Pamlico
where Washington gets its drinking water, interacted with chlo-
rine being used in the treatment plant to purify the water. The
contamination gets worse in the summer when the river water is
warm. We explained that and other technical complications to our
readers over and over again. Considering how much trouble we
had even spelling *carcinogenic* and *trihalomethanes* when we began,
we did a fair job of it.

Third, work with your local television station. That may surprise
you, but we would have done a much better job on the public service aspect of our coverage if we had. The TV station reaches residents we don't, and bringing its news staff aboard the story early
on would have improved its coverage and strengthened our local
support against official antagonism. We and the TV news team at
Washington's WITN-7 realized that too late for this story. But
we've begun to cooperate now on other local coverage, something
I think will benefit both our audiences where environmental dangers are concerned.

Fourth, lift your nose from the local grindstone to contemplate the state and national implications of what you have uncovered, and don't hesitate to seek help elsewhere when local officials aren't cooperating. This is particularly important on environmental stories because, here, the small newspaper has an advantage. You literally live closer to your environment than the editor or reporter on a large paper. Where you encounter difficulty, work the telephones. Some of our best tips and supporting information came from state and federal officials and from members of environmental groups, who often can be reached on an 800 number.

Fifth, don't wait for the too rare golden opportunity to put forth your best effort. That may seem hackneyed, but as the chronology indicates, this story broke so fast it was virtually over before we had a chance to sit back and grasp its full significance. You may not have the manpower or the financial resources of the *Philadelphia Inquirer* or the *Los Angeles Times*, but you can strive for the same professional levels with the people you do have. A corollary of this rule, in fact, is not to underestimate what your staff can accomplish. Betty Gray had only a few months of journalism experience, and Mike Voss, with the paper for 4 years, is not a college graduate. Those two, combining very different but complementary, even synergistic, talents, turned out top-quality work. So did News Editor Michael Adams, who laid out the coverage. After we won the Pulitzer, one of the comments I was proudest of came from Madelyn Ross, managing editor of the *Pittsburgh Press* and a member of the nominating committee. While the committee was aware of our limited resources, she said, the paper's size was not a significant factor in selecting it as one of the four finalists. What put us over the top, she said, was "an outstanding piece of journalism."

Finally, be prepared for surprises, both pleasant and unpleasant. One man I didn't recognize stopped his car in the middle of the street and shouted, "Well done!" Perhaps because of the large number of letters praising our coverage, I was unprepared for the antagonism of others. No enthusiasm developed when a civic banquet honoring the newspaper was proposed, and the idea was dropped. "Daily News Won Few Friends, But May Snare Pulitzer," said a *Charlotte Observer* headline.

NOW THAT WE'VE developed the sources, we plan to keep digging. We found during coverage of the water scandal that no cancer statistics were readily available to assess the damage that had been inflicted. We kept after them, and Gray was able to report in spring 1990 that Beaufort County had a death rate 25 percent higher than the rest of the state and 50 percent higher than the national average. The University of North Carolina has asked the National Cancer Institute for a grant to study Beaufort County, citing the history of Washington's water contamination and previous air pollution (now under control) from the phosphate plant.

Since the water contamination story, Gray has broken several statewide environmental exclusives, and Voss beat larger papers in the state to the story of corruption in the law enforcement arm of the N.C. Wildlife Resources Commission. "What rabbit is that [expletive] little paper down there going to pull out of the hat next?" the state editor of one of North Carolina's largest newspapers asked recently. He'll see.

GERRY STOVER

Media, Minorities and the Group of Ten

Several members of civil rights and minority groups have written to eight major national environmental organizations charging them with racism in their hiring practices.
—Philip Shabecoff, *New York Times*

Environmental justice is a fundamental human right.
—The Reverend Jesse Jackson, Rainbow Coalition

We [blacks] now have a window of opportunity to broaden the base of the environmental movement. . . . Since we're the ones being poisoned, it's really about survival.
—Bob Bullard, University of California

I don't think anybody is as aware of the whiteness of the green movement as those of us who are trying to do something about it.
—Jay D. Hair, National Wildlife Federation

Cityscape by Charles Pratt from The Sense of Wonder *by Rachel Carson,*
1965
© 1956 Rachel Carson/© 1984 Roger Christie/photo © 1965 Charles Pratt.
Reprinted by permission of Frances Collin, Literary Agent

It is the general mission of organizations within the environmental community to promote the protection, preservation and conservation of our planet's natural and environmental resources. To accomplish that goal, this community needs a commitment from a very large and diverse spectrum of our society.

A central problem faced by all major environmental and conservation organizations in the United States today is the recruitment and retention of minorities for middle, senior and executive level positions. Minorities make up a large and growing part of American society, and, indeed, the Asian and Hispanic segments of our communities are touted as our fastest-growing populations. In some communities they will outnumber the white majority by the turn of the century, when nationwide only one in four workers will be a white male.

Urban minority populations are often the first to experience the adverse physical and economic effects of environmental problems. Yet they are barely represented in the memberships, on the staffs and on the boards of environmental organizations. These organizations have been publicly criticized for racially discriminatory hiring policies and practices and their general lack of attention to those environmental issues being faced by low-income peoples and communities of color. A Feb. 1, 1990, article in the *New York Times* by Phil Shabecoff reported on demands made by a group of social justice activists to hire more minorities and address "racist hiring practices" in the environmental community.

Through the years there have been repeated attempts to increase affirmative action efforts in environmental organizations, yet, for the most part, the results have been dismal. In the past 5 years, less than 2 percent of all field staff and management hires within the environmental community have been minorities (based on an informal survey done last fall). The most commonly heard reasons for this situation are 1) "the pool of qualified candidates is almost nonexistent," or 2) "recruiting takes time under normal circumstances—recruiting minorities can be an endless task with limited or no results," or 3) "qualified minority candidates usually turn us down because of our salary levels or because of a lack of understanding of career potential within our organization."

The ramifications are clear. Not only is broader representation the morally correct objective, but it is critical if we are to maintain the ground we have gained in the fight for environmental quality and if we are to address the issues surrounding "environmental justice." To paraphrase Charles Lee and Benjamin Chavis of the United Church of Christ's Commission on Racial Justice, there can be no safe environment without a just environment.

IN THE UNITED STATES today, people of color and low-income peoples are disproportionately affected by environmental problems. In urban centers like Washington, D.C., air quality contributes to higher incidences of cancer and asthma. In the old and decaying edifices of Watts, Harlem, Roxbury, East St. Louis, Detroit and elsewhere, we find circumstances that dramatically heighten asbestos and lead poisoning problems for the local inhabitants, almost always poor people of color. The warehousing of toxic substances and the siting of toxic and solid waste disposal facilities in these communities are a consistent practice. In rural communities of the South and the Southwest, water quality is a burgeoning issue for African Americans, Native Americans and Hispanics. Yet even with the growing clamor for action from these communities, from their leaders and other advocates from the social justice movement, the large mainstream environmental organizations have applied few or no resources to addressing these issues. In order to understand why this is so, we must take a look at the organizations that make up the mainstream environmental community. The "Group of Ten" is as good a place to start this examination as any.

In 1988, I was invited to address a meeting of the Group of Ten. At the time, all I really knew about it was that it was composed of the CEOs of 10 of the country's self-proclaimed largest and most important environmental and conservation organizations. This group meets two to four times a year to discuss the state of the environmental movement in America, to determine where they will present united fronts, and to try to get a handle on and find solutions for a variety of common problems. The various members of the group, or "gang" as others often refer to it, were all facing a similar problem: There were pitifully few low-income people and people of color on their staffs and their boards, and so they were ill-equipped to address growing community-based and grass roots demands for greater attention to "environmental racism." In addition, the U.S. Labor Department's report "Workforce 2000" had strong implications for organizations such as theirs, where professional employees were mostly white males. Over the years they had verbally supported affirmative action but had done little to effect any change in the "color" of their organizations. They needed assistance in finding a successful solution to this recruitment problem, but they didn't understand (and in some cases still don't) why it was so difficult to recruit minorities.

In making some suggestions to them, I had to tell them about some facts of life. Their biggest problem was institutionalized racism, both perceived and actual. People of color who have had any experience with Group of Ten organizations are very quick to point out how "white" they are. Unlike other employment sectors in this country where the leadership may be white, male and middle- or upper-class, but the rank and file are more diverse, the staffs of environmental organizations are like their top executives—predominantly middle-class and upper-class white males, usually children of the elite, people who could afford to take positions with little or no pay. People of color don't generally have that luxury. With few representatives of the "underclass" on their staffs or on their boards, the Group of Ten have not focused on issues of importance to these communities. In fact, in most instances, they are unfamiliar with the issues that are important to low-income and minority peoples. Thus, when these organizations did try to recruit minorities, they found themselves without any relevant materials or issues with which to woo them.

When the Group of Ten publicizes their efforts on behalf of the environment, they use the standard advertising techniques. Posters, print and electronic media ads, and direct marketing strategies all emphasize their middle-class nature. Media research tells them that they get a lot of their financial support from that segment of society that we have come to call yuppies. Young, well-off and white, these are the people who, in their minds, go hiking and camping and value nature—likely candidates to woo for jobs and financial support. While not intentionally eliminating minorities from their applicant pools, the group has effectively accomplished just that. People of color don't see a place for themselves in this movement: There is no indication that the work these organizations are doing will do anything to improve the declining environmental conditions of their communities. If anything, minorities see the Gang of Ten as more closely aligned with the American Establishment—which, having employed the economic blackmail of trading jobs for the environment, was often responsible for local environmental degradation in the first place—than with the causes of environmental and social justice. To top it off, qualified minorities are a high-priced commodity on the open market. The Group of Ten offer them little or no professional inducement, and the salary levels generally available to the few minorities who do apply are laughable compared with what they can get elsewhere in the corporate sector.

To diversify the environmental mainstream, the Group of Ten and other environmental organizations must change the way they do business.

First, they must embrace and support social justice *and* civil rights issues. In minority and low-income communities, environmental issues are often synonymous with issues of race and class. The siting of toxic waste dumps is more than just an "environmental" issue to the communities that have to live with them. It branches out to issues of health, jobs and job displacement, and it is directly related to the economic and political empowerment issues of institutional racism. The same is true for air quality issues in cities and water quality issues in rural communities. The mainstream environmental community must declare its support of the goals of equality across the board, and then they can expect to begin having the support of low-income peoples and people of color.

Second, the environmental community must take the time to learn about the needs and priorities of people of color. This means becoming sensitized to the cultural and ethnic differences between groups of people and then learning to be a true partner in addressing issues, as opposed to making decisions and determining the agenda for everyone else. Patronizing attitudes that have historically surfaced in previous attempts at joint efforts have gone a long way toward alienating the social justice and civil rights communities. By recognizing that there may be different perspectives on various issues and by being willing to approach solutions that are more appropriate for other constituencies than their traditional ones, the environmental mainstream will begin to have credibility with people of color.

There must be active and targeted recruitment of minorities from both the community at large and from the organizations that comprise the grass roots and social justice movement. Since many of these organizations have evolved into highly technical organisms that require specially trained and/or educated staff, they must begin to institute programs that will provide the necessary training and education to this new constituency and that will allow them to be productive within these organizations. Developing ongoing relationships with community colleges; historically black colleges and universities; tribally controlled academic institutions for Native Americans; black churches; and other minority, community-based organizations will help in creating programs to train a future work force for the environmental community.

The Group of Ten and their sister organizations must also begin to address the media issue. Why, when the media are constantly bombarding us with the message that the '90s is the decade of the environment, do we not hear about the environmental degradation occurring in "Cancer Alley" in Louisiana, or the fact that lead poisoning is fast becoming the primary culprit in escalating incidences of birth defects and infant mortality in poor urban communities? Why aren't Americans outraged over the poor air quality that contributes to respiratory illnesses and skin cancers among the inhabitants of industrialized inner cities? When can we expect Ted Koppel or Jane Pauley or "60 Minutes" to focus public attention on the blatant racism involved in the siting of hazardous landfills and toxic waste dumps in low-income and minority communities? Is it more important to save a baby seal, an elephant or a whale than it is to save the life of a child of color?

In this country the media are the primary forgers of public opinion and thus the primary motivators of public action. The environmental mainstream has, over the years, become very sophisticated in its manipulation of the media for its benefit. By incorporating minority issues into its agendas and by applying the same attention and resources to them that it has in the past to its conservation-oriented goals, the environmental movement can begin to show the media and the public a more representative face. The Group of Ten, too, will benefit from this increased exposure. The *Chronicle of Philanthropy* points out in a recent article that minorities, particularly African-Americans, give proportionately more to charitable causes. As minorities begin to see the issues that affect their communities gaining more attention, particularly from the mainstream environmental community, they will support those efforts. They will become members, they will send donations, and they will volunteer their time.

The media have an obligation of their own to educate the public on these issues. Articles by courageous journalists like Phil Shabecoff in the *New York Times* and Paul Ruffins in the *Los Angeles Times* and the *Washington Post* have begun to appear. Yet the majority of the public discussion of this subject has taken place in regional journals and newsletters or in the publications coming out of the social justice and civil rights communities such as the *New Age Journal*, the *Utne Reader* or *Race, Poverty and the Environment*. The mainstream media should take a lesson from their more progressive brethren. It is the job of journalists and the media to report on items and issues of importance and concern to *all* of us, not just those special interests that can afford to pay for the attention. It is also the responsibility of the media to encourage justice in our society regardless of the color of victims. I contend that the American media have failed in this responsibility.

Ultimately, the mainstream environmental community and the media are going to have to recognize that there is a ground swell occurring among low-income peoples, communities of color, grass roots groups, and social justice and civil rights organizations. Change *is* going to come to the face of this nation's environmental movement, and hence to the priorities we place on the resources we use to clean up and protect our human environment. It behooves the Group of Ten and its sister organizations to join this movement, and it is my hope—indeed my expectation—that the American media and its journalists will be a catalyst in effecting this change and will chronicle it for the benefit of us all.

The dustbowl: Dallas, South Dakota, 1936.
U.S. Soil Conservation Service (114-SD-5089)

DANA JACKSON

Who Speaks for the Land?

On Monday morning, March 12, 1991, the announcer on radio station KHCC in Hutchinson, Kansas, advised listeners who might be planning to drive through the western part of the state to change their minds and stay home. The station predicted an imminent dust storm, with complete blackouts, zero visibility on the highways, and winds gusting up to 70 miles an hour—strong enough to topple tall vehicles.

The storm threatened not only drivers but also the winter wheat crop. Topsoil moisture was low, and an early spring had brought the plants out of dormancy. The winds sandblasted them, sucking them dry.

All day the winds blew over thousands of acres of wheat, scooping up the soil between the long rows and from the bare spots where there was a "poor stand." Both the dust storm and the circumstances attending it hauntingly recalled the dust-bowl years of the Great Depression; more than a few Kansans wondered if a new dust bowl were in the making.

The great dust storms of the 1930s followed the great plowing of the Plains, the "busting" and "breaking" of the prairie. Prior to the coming of the plow, even in drought years deep-rooted grasses had hung on to the soil during wind storms; but commercial farming for maximum production in the expansionist, laissez-faire economic climate of the 1920s, as much as the drought, caused the dust bowl. Farmers did not think about the environmental limits of that great expanse of land; to them it was a natural resource waiting to be used. But when dust from the Great Plains settled over furniture in Boston, New York and Washington, Congress began to think otherwise, and in 1936 it established the Soil Conservation Service, encouraging farmers to use soil resources more wisely.

In most people's minds, soil loss is a thing of the past, yet by the 1970s the rates of soil erosion in the United States were higher than they had been even in the 1930s. The postwar rise of mechanization and chemicalization enabled farmers to produce high yields even when soil eroded up to 20 or 30 tons per acre per year. Farmers, urged to produce ever higher yields, tilled over terraces and broke them down, took out windbreaks and shelterbelts, and farmed one crop for years in the same field. The new synthetic fertilizers could make up for lost fertility when soil blew or washed away, and new pesticides successfully controlled insects. In the mid-1970s, then Secretary of Agriculture Earl Butz was encouraging farmers to plant "fencerow to fencerow."

Today agriculture rarely makes the front pages as an environmental concern, and when it does, it is rarely in the context of "agribusiness," the engine that drives so many American farms. Instead, the media have focused largely on the potential danger to human health through consumers' ingestion of pesticides or other agricultural chemicals in food or water.

In February 1989, for example, the Natural Resources Defense Council (NRDC) released its study *Intolerable Risk: Pesticides in Our Children's Food*, which estimated the potential health risks to children ages 1 to 5 from dietary exposure to 23 pesticides resulting from consumption of 27 fruits and vegetables—and in particular, the dangers of children developing cancer following early exposure to Alar, a chemical used in 1988 on 15 percent of the nation's apple orchards to make the fruit ripen at the same time. Although the NRDC and the Environmental Protection Agency (EPA) disagreed on the level of the risk, both agreed it was too high; the EPA had already decided to propose a ban on Alar when the NRDC released and publicized its study.

The effects were dramatic. After a "60 Minutes" segment on Alar and much coverage in the print and broadcast media, schools throughout the country withdrew apples from lunch menus, and the "apple a day to keep the doctor away" became more like the poisoned apple of Snow White. Apple growers who couldn't sell their produce were virtually wiped out, and they have collectively sued the NRDC for damages. The publicity also created a demand for organic produce that triggered overproduction by small organic growers and the entry of highly competitive large-scale farmers into the organic market. After a glutted market and low prices in 1990, many small organic farms went bankrupt.

But the "Alar scare" did open an intense debate about acceptable and unacceptable environmental risk. People became more aware of farmers' great dependence upon chemicals, not only to protect fruits and vegetables from insects, but, as in the case of Alar, to market them more effectively.

Often such stories tend to make it seem as if, once some isolated problem has been fixed (by regulating a "safe" residue level), we can quit being concerned. But Terry Shistar, president of the board of directors of the National Coalition Against the Misuse of Pesticides, points out that when agriculture made the choice to use pesticides, it chose a route with environmental risks all along the way. This includes hazards to workers who manufacture pesticides, the potential danger to residents near plants that manufacture pesticides (remember Bhopal?), the risks to those who mix and apply the pesticides, the contamination of surface and groundwater, and the problem of disposing of empty pesticide containers, as well as the risk to consumers from pesticide residues on food products.

But even if these issues get media attention, longer-term and systemic problems with agricultural chemicals in the environment frequently do not. Instead they are reduced to public relations battles in which the media are often unwittingly whipsawed by people who know the right buttons to push.

"Objectivity" is one of them. When the National Academy of Sciences released its study *Alternative Agriculture*, documenting the success of farmers who use few or no chemical pesticides, major newspapers throughout the country covered the report and commented on it favorably. This triggered the defenses of the agrichemical industry, which set up ACRE (Alliance for a Clean Rural Environment), an "educational program" intended to counter the report. ACRE fact sheets take the high road of scientific objectivity, stressing that until there is sound scientific evidence of health risk from pesticides, farmers should not be denied the economic benefits of their use. ACRE does stress careful use, in sort of a feigned stewardship language, and offers a palliative perspective that "it's not the technology that is the problem; it's how we use it."

Personality is another media vulnerability. Last year California voters rejected two environmental propositions on the state ballot, "Big Green" and "Forests Forever." Chemical companies contributed heavily toward the total $17 million spent to defeat the propositions, though most of the country knew only that California Congressman Tom Hayden—Jane Fonda's ex-husband and a legendary liberal—was one of the propositions' sponsors.

RACHEL CARSON'S *Silent Spring* first exposed the price of agricultural success and its dependency upon pesticides and herbicides. Carson described how persistent poisons such as DDT, chlordane, heptachlor and others—which at that time were essentially unregulated—affected species other than insects, including humans. She did not contend that all chemical pesticides should be banned, but she objected to chemicals being put "into the hands of persons largely ignorant of their potential for harm" and subjecting "enormous numbers of people to contact with these poisons without their consent and often without their knowledge." At another level, however, Carson argued that the profligate spraying of dangerous chemicals for the sake of maintaining high production simply did not make sense—that the main problem in agriculture was *over*production. *Silent Spring* startled the public, but it infuriated agribusiness and university agricultural researchers, who unsuccessfully tried to discredit the book's author.

Since *Silent Spring*, media stories about pesticides have been common. Few people would feel comfortable eating fruit if told that a chemical residue remained on it; few would accept aerial spraying of fruit trees or fields close to their houses without questioning the safety of the operation.

Why, then, does agriculture still rely so heavily on these chemicals that the public now recognizes as hazardous to our food and water? Pesticide use in the United States increased twentyfold between 1950 and 1985, and doubled between 1964 and 1985. Fertilizer use increased fourfold from 1950 to 1985.

One reason is the high-production imperative for American agriculture promoted by the U.S. Department of Agriculture (USDA). To get high yields, USDA extension agronomists advise the use of chemicals, and chemical companies advise farmers and extension agents on how to use their products. They also advertise intensely on television and in farm magazines and are a powerful influence in both Congress and agricultural research universities. Today a farmer who decides not to use chemicals risks losing his farm: Banks have refused loans to farmers unwilling to apply chemicals in pursuit of high yields.

The EPA regulates pesticides under the Federal Insecticide, Fungicide and Rodenticide Act (FIFRA). Originally enacted in 1947 to ensure pesticides efficacy—but not safety—FIFRA registration was revised after *Silent Spring* to include a concern for safety. But the law has been more effective in protecting manufacturers than the public. Because pesticides have been judged to be necessary agricultural poisons—FIFRA is based on the assumption that the only way to control pests is with pesticides—the government weighs their environmental and health risks against presumed economic benefits.

But what benefits? To whom? Studies by David Pimentel at Cornell University show that while insecticide use has increased tenfold since the 1940s, crop losses to insects, which have developed resistance to one chemical after another, have doubled during the same period. Needless to say, this means the chemical industry must continually invent new pesticides, and as predators of insect pests are killed by spraying and the ecology of fields is altered, many farmers are trapped into continuing the application of insecticides. The ironic result is overproduction and lower prices—forcing farmers to strive for even higher yields to make up for them. The federal government, meanwhile, complains about paying price supports because of overproduction even as it depends upon the export of surplus agricultural commodities to offset the balance-of-trade deficit.

Doubtless, direct poisoning and cancer risk from pesticides will continue to make the news as long as American agriculture depends so heavily on their use. But there is another "environmental" story here, largely untold, and it concerns the relationship between the land, the water and agribusiness.

FOR YEARS NOW, farm organizations, the USDA and agricultural suppliers have encouraged farmers to think of themselves as agribusinessmen. Those who raise livestock are in the livestock "industry"; those raising grapes are in the grape "industry." Presumably this is more dignified than being a mere "farmer," but it also changes the relationship between the land and those who work it. Soil and water become resources to be mined; land becomes just a commodity. The farmer is taught to adopt scientific farming practices and eliminate sentimentality from tasks such as planting fields or milking cows. As in manufacturing, the single goal of high production directs decision-making. The farmer must control nature to achieve this goal, so he or she buys those products that will help: pesticides, herbicides, large farm machinery, irrigation equipment. As the scale of the farming operation increases, the agribusinessman's connection to the land, the crops and livestock weakens.

But when the farm becomes a factory, utilizing natural resources efficiently to produce large quantities of goods and in pursuit of profit, it must be regulated just like any other large business with the potential to pollute.

Large agricultural organizations defend themselves against regulation by perpetuating certain prejudices and misconceptions about farmers and farming. The litany is a familiar one—"farmers are independent and don't need to be told how to farm by the government; they know best what is good for their land; farmers themselves wouldn't use chemicals if they were really dangerous; farmers following 'best management practices' do not create any environmental problems; U.S. agriculture feeds the world, and so farmers must use the most modern, efficient technology to keep production high; consumers and city people should not be involved in making policy that affects farmers"—and so on, right down to "environmentalists are going to make it impossible to farm." At an Iowa meeting of soil conservation district commissioners, I heard the national president of this organization describe environmentalists in Washington as enemies of farmers— demagogues who "won't be happy until big government controls everything farmers do."

News coverage of the debate over the 1990 Farm Bill exacerbated the antagonism between farmers and environmentalists and did little to further public understanding of the issues involved. Reporters covering the bill tended to contrast the views of agribusiness with those of established "Group of Ten" environmental groups, rather than provide good information about a particular issue or explore alternative views. Proponents of "sustainable" agriculture, which requires more time and labor and is especially well suited to the midsize, diversified family farm, received almost no media attention, and, needless to say, neither did their principles—crop rotation, ridge tillage, contour farming, composting, manure and legumes for fertilizer, intercropping, and pest management based on ecological understandings.

Chemical companies lobbied hard against the low-input sustainable agriculture provisions in the Farm Bill (for the obvious reason that if farmers learn to farm without herbicides, the companies will not earn a good return on their research investment), and the current political climate in Washington favors the companies. The Bush administration is committed to eliminating subsidies for farmers, and although Congress has been unwilling to go that far, it did concur in the 1990 Farm Bill on a complicated triple-base system that would decrease subsidies. Eliminating subsidies would unleash a dog-eat-dog competition in which the most efficient producers would control the market, thus accelerating the already dizzying cycle of chemicals/excess production/ lower prices/more chemicals/more excess. According to former Agriculture Secretary Clayton Yeutter and the Bush administration, such unimpeded "competition" and economic "growth" are to be valued over all other social, environmental or cultural values.

To their credit, the news media gave the 1990 Farm Bill (which, stacked, stands more than 6 feet high) unusual attention, most of it focused on the bill's environmental aspects. At the same time, however, almost every news outlet in the nation missed or ignored what could have been the biggest agriculture/environment story of the decade—the General Agreements on Trade and Tariffs (GATT).

IN JULY 1990, U.S. trade representatives took their free-market ideology to the Geneva round of the GATT talks. In a discussion paper on tariffication, the United States proposed (1) eliminating all import restrictions by converting existing import regulations to fixed tariffs and then phasing them out over time, (2) subjecting health, safety and environmental standards to GATT rules and regulations, (3) prohibiting food export restrictions, even in times of critical shortages, and (4) phasing out other farm programs, including price supports and supply management. All of these proposals would have had seriously detrimental environmental impacts had the talks succeeded, though their potential in that regard received almost no news coverage.

The first proposal would have devastated local farmers in many Third World countries who would have found themselves unable to compete against imports dumped into their markets by the United States, Australia or the European Community (EC). Farmers would have been forced to push their land to the limits to increase production, a process in which many small farms would invariably have failed and been absorbed by large-scale farms, operations that often employ the most energy- and chemical-intensive methods.

The second proposal—which received virtually no mainstream media attention—would have virtually overturned food safety and environmental regulations in the United States and elsewhere. In a plan designed by Daniel Amstutz (former senior vice president of Cargill International, the largest grain company in the world), the final authority over environmental and food safety regulations would have been taken away from national legislatures and turned over to a Rome-based U.N. advisory agency, the Codex Alimentarius. Codex currently has standards for 133 pesticides, compared with the 350 of the EPA. Sixteen percent of Codex's pesticide regulations are lower than the EPA's; Codex, for example, allows food to contain 50 times more DDT than is permitted under U.S. law. Under the concept of "harmonization," state and federal governments would have had to set their standards to the Codex level in order to remove them as non-tariff "trade barriers."

Under Codex rules, any new chemical standards would have had to be set according to "sound science"—which means simply that hazards from chemicals must be proven, not merely suspected, and in any case that proof of safety is not required. Proponents of the U.S. plan claimed that Codex standards could be raised to the U.S. level. These claims, while intended to allay fears, are entirely unrealistic: The purpose of harmonization is to *remove* trade barriers by setting a ceiling on environmental standards. It is hardly likely that any regulations would become more stringent.

The third proposal in the U.S. plan, which denied nations the right to restrict food exports even in times of famine, would almost certainly have resulted in the destruction of the environment by desperately hungry people. In the United States, where environmentalists claim uncontrolled log exports have led to the log shortage that now threatens old-growth forests in the Pacific Northwest, the U.S. GATT proposals would have forbidden limits on log exports *and* protection of the forests as restrictions to free trade.

The fourth proposal, to reduce or eliminate domestic farm support programs, could have resulted in the elimination of a wide variety of environmental protection and conservation programs—from the conservation reserve program that pays farmers to take highly erodible land out of production and plant it in grass or trees, to economic incentives to protect water quality through sustainable farming techniques.

Those conservation programs also serve as supply management programs to avoid overproduction. But under the GATT free-trade agreements, supply management would have been illegal, and the lowering of world commodity prices which would have resulted would have put pressure on natural resources. Farmers all over the world would have been forced to intensify their production to make up in volume what they would have lost in lower prices. They would have done it by using more fertilizers, pesticides and large equipment, or by opening up more land, probably marginal and highly erodible, for production. Who would have benefited from such free trade? Chemical and machinery manufacturers.

Such unfettered exploitation of natural resources by international agricultural conglomerates is an example of what David Brower has called "strength through exhaustion." It provides short-term gain, but at the price of international and intergenerational justice.

Stories in the U.S. media about the GATT talks mostly followed news releases by the federal government, and with few exceptions they dealt with the subsidy dispute, not the environmental issues. In Europe the media reported on large demonstrations by farmers objecting to the cessation of farm subsidies. The talks broke down in December due to European delegates' refusing to agree to the 75 percent cuts in subsidies that the United States had demanded, though even the 30 percent reductions that the Europeans proposed would have seriously affected the number of farms. The United States lost 250,000 farms during the farm crisis of the '80s, while the EC has been losing farmers at the rate of 250,000 *annually*.

One of the most disturbing aspects of the GATT process was the lack of participation by citizens. In the United States there was no public debate about the social and environmental impact of the proposals prior to the U.S. participation, especially worrisome since under the GATT fast-track system Congress would have had only 90 days to approve or reject the entire GATT package—with no amendments.

How could a free and democratic country subject itself to such a process, surrendering the power of choice to a superagency? Why would our own government promote such environmentally devastating international trade policies? The most obvious and important reason, of course, is that Americans were not very well informed about the GATT agreements and the process of approving them.

The most informed and active observers in the GATT process were those hoping to achieve economic advantage if the U.S. proposals were approved. In one official U.S. trade delegation, 12 out of 28 representatives were executives from transnational food-processing companies and associations, among them Coca-Cola, Kraft, Ralston Purina, the Grocery Manufacturers of America and the American Frozen Food Institute. At a meeting of the

Codex Commission, the U.S. delegation included executives from chemical giants DuPont, Monsanto and Hercules. A USDA official described the relationship between industry and government representatives as "very close to a collegial atmosphere."

Delegates to the GATT talks reopened negotiations in March 1991. Their outcome is uncertain at this writing, but the involvement of multinational grain companies, chemical companies and food-processing industries in negotiations does not bode well for the family farm or environmental protection.

ENVIRONMENTAL PROBLEMS IN agriculture, whether they be directly endangering consumer health or degrading natural resources over the long term, cannot be understood or solved without understanding that most are the consequence of industrialization applied inappropriately to biological systems. Neither can they be understood by an ethic that champions competition in the world marketplace as the only test of who is best suited to produce and process food for human consumption.

The solutions to environmental problems in agriculture require not only changing these assumptions, but changing them radically—developing an understanding that the natural resource base for agriculture is biological. Good farming is an art or a craft as much as it is a business, and it can be practiced sustainably only within a cultural and social context that relies on renewable resources rather than nonrenewable ones. For a generation now, the media have informed the consuming public about immediate threats to their health, but they must begin to examine and explain the long-term, systemic environmental problems that precede them. The consequences of industrial production methods on the long-term ability of the land to produce food are far more serious than current coverage suggests. The health and well-being of generations to come depend upon the availability of good-quality soil and water, the major resources in agriculture. If we do not pay attention to the effects of high-production agricultural technology upon the land, if we do not curb the growth imperative, we will destroy options for future generations.

ECONOMICS
and
ENVIRONMENTAL
POLICY-MAKING

Ferdinand the Contented Bull
From The Story of Ferdinand *by Munro Leaf, drawings by Robert Lawson.*
Copyright 1936 by Munro Leaf and Robert Lawson, renewed © 1964 by
Munro Leaf and John W. Boyd. Used by permission of Viking Penguin, a
division of Penguin Books USA Inc.

HERMAN E. DALY

Boundless Bull*

If you want to know what is wrong with the American economy, it is not enough to go to graduate school, read books and study statistical trends—you also have to watch television. Not the Sunday morning talking-head shows or even documentaries, and especially not the network news, but the really serious stuff—the commercials. For instance, by far the most penetrating insight into the American economy as of the late '80s and early '90s is contained in the image of the bull that trots unimpeded through countless Merrill Lynch commercials.

One such ad opens with a bull trotting along a beach. He is a very powerful animal—nothing is likely to stop him. And since the beach is empty as far as the eye can see, there is nothing that could even slow him down. A chorus in the background intones: "to . . . know . . . no . . . boundaries. . . ." The bull trots off into the sunset.

Abruptly the scene shifts. The bull is now trotting across a bridge that spans a deep gorge. There are no bicycles, cars or 18-

---

*The views presented here are those of the author and should in no way be attributed to the World Bank.

wheel trucks on the bridge, so again the bull is alone in an empty and unobstructed world. The chasm, which might have proved a barrier to the bull, who after all is not a mountain goat, is conveniently spanned by an empty bridge.

Next the bull finds himself in a forest of giant redwoods, looking just a bit lost as he tramples the underbrush. The camera zooms up the trunk of a giant redwood whose top disappears into the shimmering sun. The chorus chirps on about a "world with no boundaries."

Finally we see the bull silhouetted against a burgundy sunset, standing in solitary majesty atop a mesa overlooking a great, empty southwestern desert. The silhouette clearly outlines the animal's genitalia, making it obvious even to city slickers that this is a bull, not a cow. Fade-out. The bull cult of ancient Crete and the Indus Valley, in which the bull god symbolized the virile principle of generation and invincible force, is alive and well on Wall Street.

The message is clear: Merrill Lynch wants to put you into an individualistic, macho world without limits—the U.S. economy. The bull, of course, also symbolizes rising stock prices and unlimited optimism, which is ultimately based on this vision of an empty world where strong, solitary individuals have free reign. This vision is what is most fundamentally wrong with the American economy. In addition to television commercials it can be found in politicians' speeches, in economic textbooks and between the ears of most economists and business journalists.

No bigger lie can be imagined. The world is not empty; it is full! Even where it is empty of people, it is full of other things. In California it is so full that people shoot each other because freeway space is scarce. A few years ago they were shooting each other because gasoline was scarce. Reducing the gasoline shortage just aggravated the space shortage on the freeways.

Many species are driven to extinction each year due to takeover of their "empty" habitat. Indigenous peoples are relocated to make way for dams and highways through "empty" jungles. The "empty" atmosphere is dangerously full of carbon dioxide and pollutants that fall as acid rain.

Unlike Merrill Lynch's bull, most do not trot freely along empty beaches. Most are castrated and live their short lives as steers imprisoned in crowded, stinking feedlots. Like the steers, we too live in a world of imploding fullness. The bonds of community, both moral and biophysical, are stretched, or rather compressed, to the breaking point. We have a massive foreign trade deficit, a domestic federal deficit, unemployment, declining real wages and inflation. Large accumulated debts, both foreign and domestic, are being used to finance consumption, not investment. Foreign ownership of the U.S. economy is increasing, and soon domestic control over national economic life will decrease.

Why does Merrill Lynch (and the media and academia and the politicians) regale us with this "boundless bull"? Do they believe it? Why do they want you to believe it, or at least to be influenced by it at a subconscious level? Because what they are selling is growth, and growth requires empty space to grow into. Solitary bulls don't have to share the world with other creatures, and neither do you! Growth means that what you get from your bullish investments does not come at anyone else's expense. In a world with no boundaries the poor can get richer while the rich get richer even faster. Our politicians find the boundless bull cult irresistible.

The boundless bull of unlimited growth appears in economics textbooks with less colorful imagery but greater precision. Economists abstract from natural resources because they do not consider them scarce, or because they think that they can be perfectly substituted by man-made capital. Either the natural world puts no obstacles in the bull's path or, if obstacles like the chasm appear, capital (the bridge) effectively removes them.

Economics textbooks also assume that wants are unlimited. Merrill Lynch's boundless bull is always on the move. What if, like Ferdinand, he were to just sit, smell the flowers, and be content with the world as it is without trampling it underfoot? That would not do. If you are selling continual growth, then you have to sell continual, restless, trotting dissatisfaction with the world as it is, as well as the notion that it has no boundaries.

This preanalytic vision colors the analysis even of good economists, and many people never get beyond the boundless bull scenario. Certainly the media have not. Would it be asking too much of the media to do what professional economists have failed to do? Probably so, but all disciplines badly need external critics, and in the universities disciplines do not criticize each other. Even philosophy, which historically was the critic of the separate disciplines, has abdicated that role. Who is left? Economist Joan Robinson put it well many years ago when she noted that economists have run off to hide in thickets of algebra and left the really serious problems of economic policy to be handled by journalists. Is it to the media that we must turn for disciplinary criticism, for new analytic thinking about the economy? The thought does not inspire confidence. But in the land of the blind, the one-eyed man is king. If journalists are to criticize the disciplinary orthodoxy of economic growth, they will need both the energy provided by moral outrage and the clarity of thought provided by some basic analytic distinctions.

Moral outrage should result from the dawning realization that we are destroying the capacity of the earth to support life and counting it as progress, or at best as the inevitable cost of progress. "Progress" evidently means converting as much as possible of Creation into ourselves and our furniture. "Ourselves" means, concretely, the unjust combination of overpopulated slums and overconsuming suburbs. Since we do not have the courage to face up to sharing and population control as the solution to injustice, we pretend that further growth will make the poor better off instead of simply making the rich richer. The wholesale extinctions of other species, and some primitive cultures within our own species, are not reckoned as costs. The intrinsic value of other species, their own capacity to enjoy life, is not admitted at all in economics, and their instrumental value as providers of ecological life-support services to humans is only dimly perceived. Costs and benefits to future humans are routinely discounted at 10 percent, meaning that each dollar of cost or benefit 50 years in the future is valued at less than a penny today.

But just getting angry is not sufficient. Doing something requires clear thinking, and clear thinking requires calling different things by different names. The most important analytic distinction comes straight from the dictionary definitions of growth and development. *To grow* means to increase in size by the accretion or assimilation of material. *Growth* therefore means a quantitative increase in the scale of the physical dimensions of the economy. *To develop* means to expand or realize the potentialities of; to bring gradually to a fuller, greater or better state. *Development* therefore means the qualitative improvement in the structure, design and composition of the physical stocks of wealth that results from greater knowledge, both of technique and of purpose. A growing economy is getting bigger; a developing economy is getting better. An economy can therefore develop without growing, or grow without developing. A steady-state economy is one that does not grow, but is free to develop. It is not static—births replace deaths and production replaces depreciation, so that stocks of wealth and people are continually renewed and even improved, although neither is growing. Consider a steady-state library. Its stock of books is constant but not static. As a book becomes worn out or obsolete, it is replaced by a new or better one. The quality of the library improves, but its physical stock of books does not grow. The library develops without growing. Likewise the economy's physical stock of people and artifacts can develop without growing.

The advantage of defining *growth* in terms of change in physical scale of the economy is that it forces us to think about the effects of a change in scale and directs attention to the concept of an ecologically sustainable scale, or perhaps even of an optimal scale. The scale of the economy is the product of population times per-capita resource use, i.e., the total flow of resources—a flow that might conceivably be ecologically unsustainable, especially in a finite world that is not empty.

The notion of an optimal scale for an activity is the very heart of microeconomics. For every activity, be it eating ice cream or making shoes, there is a cost function and a benefit function, and the rule is to increase the scale of the activity up to the point where rising marginal cost equals falling marginal benefit, i.e., to where the desire for another ice cream is equal to the desire to keep the money for something else, or the extra cost of making another pair of shoes is just equal to the extra revenue from selling the shoes. Yet for the macro level, the aggregate of all microeconomic activities (shoe making, ice cream eating and everything else), there is no concept of an optimal scale. The notion that the macro economy could become too large relative to the ecosystem is simply absent from macroeconomic theory. The macro economy is supposed to grow forever. Since GNP adds costs and benefits together instead of comparing them at the margin, we have no macro-level accounting by which an optimal scale could be identified. Beyond a certain scale, growth begins to destroy more values than it creates—economic growth gives way to an era of anti-economic growth. But GNP keeps rising, giving us no clue as to whether we have passed that critical point!

The apt image for the U.S. economy, then, is not the boundless bull on the empty beach, but the proverbial bull in the china shop. The boundless bull is too big and clumsy relative to its delicate environment. Why must it keep growing when it is already destroying more than its extra mass is worth?

Because (1) we fail to distinguish growth from development, and we classify all scale expansion as "economic growth" without even recognizing the possibility of "anti-economic growth," i.e., growth that costs us more than it is worth at the margin; (2) we refuse to fight poverty by redistribution and sharing, or by controlling our own numbers, leaving "economic" growth as the only acceptable cure for poverty. But once we are beyond the optimal scale and growth makes us poorer rather than richer, even that reason becomes absurd.

Sharing, population control and true qualitative development are difficult. They are also collective virtues that for the most part cannot be attained by individual action and that do not easily give rise to increased opportunities for private profit. The boundless bull is much easier to sell, and profitable at least to some while the illusion lasts. But further growth has become destructive of community, the environment and the common good. If the media can help economists and politicians to see that, or at least to entertain the possibility that such a thing might be true, they will have rendered a service far greater than all the reporting of statistics on GNP growth, Dow Jones indexes and junk bond prices from now until the end of time.

Yellowstone Lake: William Henry Jackson, 1872. Photographs by Jackson and others were critical in securing congressional support for the preservation of public lands and the establishment of several national parks.

EMILY T. SMITH

Greens and Greenbacks

In *Cabaret*, Master of Ceremonies, the lead character of the musical set in Berlin's demimonde during the 1930s, belts out the creed of his era in a song: "Money makes the world go round, world go round." The axiom is as apropos for environmental issues—and coverage of the environment.

If anyone doubts that money and the environment are indivisible, just look back to the weeks preceding the 20th anniversary of Earth Day on April 22, 1990. No one in the media could escape companies trumpeting their "Green" deeds. A blue chip lineup of U.S. industry inundated reporters with phone calls and press releases touting everything from environmentally friendly products to wildlife conservation programs. Consumer-product giant Proctor & Gamble unveiled a line of recycled and refillable packaging. International Business Machines and American Telephone & Telegraph promised to eliminate ozone-destroying CFCs from their production processes before a deadline mandated by an international treaty to phase out the chemicals. One beleaguered journalist tallied over 200 such phone calls and collected over a thousand pages of press materials prepared for the occasion before giving up the count.

Industry couldn't afford to miss a chance to get a positive environmental story out. During the late 1980s it became clear that the environment would become an ever greater factor on a company's bottom line. The shrinking ozone hole, disappearing rainforests and global warming became hazards the public refused to ignore. A continuing saga of industrial accidents, from Chernobyl to the Exxon *Valdez*, raised public wrath. Public opinion polls showed that the majority of the public blames business most for pollution and thinks that industry isn't doing enough to clean up.

Federal and state regulators dramatically stepped up the enforcement of pollution laws, a sobering trend for U.S. industry, which by some estimates is sitting on a $150 billion cleanup tab under federal laws governing hazardous wastes and contaminated sites alone. A rash of new state and federal regulations was increasingly impinging on corporate autonomy to locate facilities, organize production processes and produce products. By 1990, in polls among members of the prestigious Business Roundtable and the World Economic Forum, business leaders ranked the environment as one of the most potent forces shaping industry and world economies in the next decade. In explaining why environmental concerns became a top priority at DuPont, chairman and CEO Edgar S. Woolard Jr. made it clear why few companies can afford to ignore the environment. "DuPont's biggest vulnerability," Woolard said, "is in the potential negatives of some environmental issues; yet its greatest opportunities lie in producing environmentally benign products and technology. Our continued existence as a leading manufacturer requires that we excel in environmental performance. . . ."

The media played a role in making the environment an industry priority in the 1980s. In their rush toward environmental coverage, the business angle was an easy and obvious one to embrace. Industrial accidents, the wages of past industrial pollution brought to light, and corporate polluters brought to task under the law routinely became grist for copy and played into the media's forte of covering conflict, crises and tragedy. Television outdid itself during the Exxon *Valdez* oil spill, reeling out horrifying images of dying, oil-soaked otters and fouled beaches. The print media followed Exxon's machinations during and after the spill in excruciating detail.

Similarly, the media jumped on the efforts of industry to go Green. DuPont's announcement that it would phase out its manufacture of chlorofluorocarbons (CFCs) in the wake of an international treaty to ban the chemicals made headlines, partly because of the $750 million price tag attached to the decision. In 1990 *Fortune* did a cover on the "Greening" of business, while the efforts of companies from Amoco, Caterpillar, General Electric and Mobil Oil in "Green marketing"—the selling of less toxic, recycled or environmentally sound products— as well as cutting waste and stopping air pollution provided the fodder for front page news and leaders on the business pages in the *Washington Post, New York Times, Wall Street Journal* and *Los Angeles Times*. Indeed, when McDonald's decided to switch from plastic to paper packaging in its restaurants and StarKist agreed to buy only tuna that had been caught with nets that didn't kill dolphins, both companies made headlines and network news.

As we move into the 1990s, however, the dollars and cents of the environment will become far more vital, and the news stories will have to become more sophisticated, more knowledgeable, more critical. As the attention of the public and policy-makers evolves from defining environmental problems and debating their severity toward taking steps to address them, the environment will be an ever more important business and economics story.

And for very good reason. When the debate shifts from whether acid rain kills trees or whether dioxin causes cancer to how to conserve forests, how to save salmon endangered by the dams that provide power to the Pacific Northwest, or how to prevent pollution and slow down the depletion of valuable resources, the influence of industry and economic and technological considerations becomes more powerful in decision-making. Industry, through the processes it uses and products it produces, *is* a major polluter.

Any regulatory efforts to mitigate or prevent pollution have implications for the costs and conduct of U.S. companies and their ability to compete abroad; at the same time, many of those companies are also the source of the technologies—whether solar power, electric cars, recycled paper, alternative fuels or biodegradable plastics—that are essential to addressing environmental problems. For these reasons, environmentalists and government alike will increasingly tap industry expertise and know-how to help set standards, determine the regulatory process, and assess technology.

But the economics of environmental issues is much bigger than industry. Efforts to protect the environment are increasingly colliding with the activities and technologies—whether it's clear-cutting forests, burning fossil fuels, or draining rivers to water fields and quench the thirst of cities—that traditionally support economic growth and provide jobs—in short, with the entire system of enterprise. Any substantial changes in that system, like a halt to the cutting of old-growth forests in the Northwest, would in the short term eliminate jobs and disrupt local and regional economies. And while alternative technologies that might be environmentally less destructive, such as solar power and energy conservation, offer the potential to create whole new industries— and jobs—they require expensive investments, and they, too, could eclipse older industries and companies, causing economic dislocation. In addition, the venue for many environmental problems has shifted from the national or regional to the global. Any effort to stave off a potential climate change, to preserve rainforests, ocean fisheries or the ozone layer, depends on global consensus. But the actions of individual nations will be shaped within the context of complex considerations of international trade, debt and national economic development goals.

In these circumstances it's nearly impossible to divorce the solutions to environmental problems from what they cost, who will pay the bill for them, and how they will impact jobs and standards of living. As Roberto L. Rapetto, senior economist at the World

Resources Institute, points out, "most conflicts over whether and how to address environmental hazards boil down to one argument: How much will it cost." Most often the players in these conflicts are national, regional or private interests who will bear the initial cost of any action and who are pitted against environmentalists and others fighting for what they contend is a larger public good.

Those were certainly paramount issues in the battle over the landmark Clean Air Act of 1990, the most sweeping piece of environmental legislation ever made law in the United States. For almost 2 years, Congress, environmentalists and industry fought over how strict to make standards for various air pollutants, what technology should be required to control them, and what regulations would best accomplish the task. The nub of the debate somehow and in some form always came down to money. Coal-burning utilities in the Midwest, for instance, bitterly opposed curbs on the emissions of acid-rain-causing sulfur dioxide, which would require them to make costly investments. Members of Congress from the Northeast, where acid rain falls, insisted upon them.

Similar considerations shape global action. By early 1991 neither India nor China had yet signed on to the Montreal Protocol, an international agreement to phase out CFCs, first signed in 1988. The reason? CFCs are used in everything from refrigerators to air conditioners, and both countries are engaged in massive industrialization efforts that include bringing modern conveniences like refrigeration to millions of their citizens. Neither nation is willing to wait for safer chemical substitutes to be developed; nor—given a shortage of hard currency, foreign debt and widespread poverty—does either have the wherewithal to pay the premium for the better technology. Their participation, which is essential if depletion of the ozone layer is to be halted, will depend on whether or not industrialized nations are willing to underwrite—at least to some extent—the cost of making available to them new, safer chemical substitutes.

The point is that the solutions to environmental problems will increasingly revolve around trade-offs between social and political goals and economic impacts. How clean is clean when it comes to detoxifying Superfund sites, for example? Depending on your standards, it could cost business anywhere from $100 billion to $300 billion, according to government and private estimates. What kind of policies will most economically foster environmental protection—pollution regulations or taxes on polluting activities? What is the best way to slash pollutants from cars—cleaner gasoline, natural gas or ethanol, which is produced from corn? (Never mind for the moment the argument over increasing cars' fuel efficiency.) Not surprisingly, the agricultural lobby supports ethanol, while the oil industry backs cleaner gas. Should loggers in the Northwest lose their jobs to preserve large tracts of old-growth forest? What kind of economic aid or incentives would be most effective in getting developing nations to protect their rainforests?

UNFORTUNATELY, AS THE environment becomes more of an economic issue, it will only exacerbate the difficulties the media confront in covering it. Deciphering the distinctions between technological options or sorting out the economic implications of environmental regulations hardly make the gut-wrenching copy or dramatic headlines that an oil spill does. Nor are the issues as "telegenic" as dying dolphins or an oil-soaked beach. More important, understanding the economics of the environment requires that reporters and editors add a new layer of knowledge about economics, business and technology on top of what they need to know about the science, scientific uncertainty and risk assessment of environmental problems.

It is difficult, for instance, to assess the relative merits and economies of alternative technologies, or to understand why companies might not adopt otherwise economical and environmentally safe practices, without understanding the role played by subsidies to industries and by other government programs or tax policies in promoting waste, resource destruction and pollution, and in determining technology choices. Agriculture, for example, is a major environmental hazard: Fertilizers and pesticides pollute water, intensive methods of cultivation erode soil, and irrigation depletes groundwater reserves. The techniques of alternative agriculture can dramatically curb erosion and chemical use without harming yields or profits, according to several reports, including one study by the National Academy of Sciences; yet this type of agriculture has made few inroads in the United States. One major reason is the federal government's commodity programs. Subsidies paid to farmers on the basis of yield and the continual production of a few crops like corn and soybeans are powerful incentives to overuse chemicals and not adopt crop rotation, a key technique of alternative agriculture.

In the future, economic actions—and economic stories—will increasingly have an environmental angle. To take but one example, the role that huge debt loads play in accelerating the destruction of rainforests, wildlife, and the destructive planting of commodity crops in developing countries will become pivotal as global efforts mount to conserve species and slow any potential global warming.

The environmental impact of trade policy is also becoming more important as trade restrictions in Europe and North America are dismantled. By early 1991, for instance, Bush administration efforts to negotiate a free-trade pact among the United States, Canada and Mexico had become, in part, an environmental issue. By then some U.S. companies had already moved over the Mexican border from Southern California in search of cheap labor and Mexico's more lenient pollution regulations. Rivers near several small Mexican border towns were running brown from the industrial pollutants dumped into them. If tariffs are ultimately removed on all goods from those countries, and Mexico doesn't adopt and enforce environmental regulations comparable to those in the United States, many more companies will flock to the area, creating a dramatic increase in air pollution and toxic waste problems. Just how those issues are debated and resolved during negotiations becomes an environmental issue.

Unfortunately, with few exceptions these economic-environmental stories have not been covered very well. Given television's reliance on the soundbite and the print media's propensity to reduce complex issues to a two-sided debate, the temptation to cover the conflict and politics that surround the economics of environmental issues—rather than the substance—is almost overwhelming. That was certainly the case in coverage of California's Proposition 128 during the fall of 1990. Known as Big Green, the initiative was without doubt the most all-encompassing piece of environmental legislation ever proposed. Among other actions, it would have banned 19 chemicals shown in some tests to cause cancer, required a 40 percent reduction in carbon dioxide emissions by the year 2010, and created a "Green czar" to watchdog the state government.

Timber, agricultural and industrial interests united in a campaign that ultimately spent $57 million to defeat the measure in the November ballot. They were pitted against environmentalists and supporters led by Hollywood celebrities and state Rep. Tom Hayden, the controversial 1960s activist and former husband of actress Jane Fonda. Both sides traded propaganda and wildly divergent estimates of what the bill would cost business interests and taxpayers. Farmers howled that it would put them out of business. Companies claimed it would raise costs and make them noncompetitive. Opponents of the measure also reeled out numbers of what it would cost consumers—in everything from gasoline to utility bills—and they played up Hayden's involvement in hopes of alienating voters. Big Green's supporters, meanwhile, used a hairless, cancer-stricken child in one commercial to underscore their contention that chemicals were a danger to the public.

For months before the ballot, California's major newspapers pretty much covered the politics and propaganda of the race for voter support, running by one count several dozen stories about the political ramifications of Hayden's involvement. It wasn't until a workshop on environmental reporting held at the *Los Angeles Times* in October that reporters at that paper realized they had failed to adequately cover the underlying issues of whether the bill itself was workable and would accomplish what was intended. After that, the paper ran a series of stories analyzing the initiative in detail. Ultimately, the *Times'* editorial page recommended a no-vote, concluding that the complexity and poor thinking underlying the bill's construction outweighed its benefits.

GIVEN THE ECONOMIC stakes involved in addressing environmental problems, and the variety of powerful lobbies with vested interests in their outcomes, it's hard to imagine what advantage there is to the public—or the media—if the media risk their credibility by adopting an advocacy position outside the editorial or op-ed pages. Good old-fashioned reporting techniques (which include a healthy dose of skepticism), computer data bases, and the environmental reporter's knowledge of science, ecology and risk will go a long way in helping journalists report on companies' efforts to develop less-polluting products and processes, on technology trade-offs, and on the validity of the claims companies make for their products.

But that may not be sufficient. As Donella Meadows suggests elsewhere in this book, a society's choices and a journalist's story are dependent upon the systems that provide information for the analysis. To adequately probe the economics of environmental solutions, or the issues that today's environmental dangers raise for economic development—and inform the public—puts new demands on reporters to examine the assumptions and information paradigm underlying current economic analysis and economics itself. For that they will need to look beyond conventional thinking to alternative visions, analyses and new ideas about the links between economics and the environment, technology, economic development and regulatory mechanisms.

Consider the all-important question of costs. If taking action on environmental problems depends in no small measure on how much it costs to do so, then the analyses used to determine those costs are critical. To meet the standards mandated by the Clean Air Act of 1990, for instance, will cost U.S. business and the economy about $40 billion a year, according to government estimates. But ask economist Rapetto, and he estimates the cost at "close to zero."

How do you account for the difference? The government uses a conventional cost-benefit analysis to get its figure, and one of the conventions of the analysis is that the environment is what economists call an externality: The beneficial services that ecosystems provide for society—such as the protection from ultraviolet light provided by the ozone layer—and the losses to society from environmental damage caused by pollution, soil erosion or the destruction of wildlife for the most part simply aren't accounted for in public and private economic decision-making. In other words, they don't count. Environmental costs and benefits are also absent from the gross national product (GNP), that all-important measure of a nation's consumption and, by inference, its wealth. Factories, equipment and office buildings are tallied as economic assets, and the decline in their value is subtracted from national wealth as they age. But most countries don't consider their forests, wildlife, clean water or soil as assets, nor do they count the destruction of these resources as a debit when calculating wealth.

These oversights are hardly academic. The economic system's blind eye to the benefits of environmental amenities and the costs of their degradation or exploitation contribute to undervaluing renewable resources such as water, timber and wildlife, according to some economists. This failure only perpetuates environmental loss, even as government subsidies for resource use and other policies accelerate it.

But this oversight has potentially damaging *economic* consequences as well. Just as business and economic reporters cannot factor out environmental factors from many stories without oversimplifying them, neither can economists safely ignore environmental costs. Indeed, by not fully considering environmental costs and benefits in determining action for environmental protection, decision makers and the public run the risk of *underestimating* the economic threat that environmental damage poses and *overestimating* the costs of addressing it, either of which could lead to ill-advised inaction on serious problems. This, then, explains how Rapetto arrived at his estimate for the cost of the Clean Air Act: When natural resources, the economic losses from air pollution, and the benefits derived from less of it are accounted for, his estimate of "almost zero" is not unreasonable.

WHEN PROBLEMS OF environmental protection are distilled into their fundamental components of what they will cost and who will pay for them, journalists could take a long step toward helping the public make informed decisions on environmental and economic trade-offs simply by questioning conventional economic analysis. When an environmental conflict gets down to a matter of dollars and cents, an alternative analysis that accounts for the environment and tallies the costs and benefits of stopping pollution or saving resources could provide valuable insight—or at the very least another perspective.

If journalists can do that, a logical and necessary next step will be to identify how economic barriers—subsidies, tax policies and the like—inhibit the development and adoption of clean technologies and the conservation of resources and how governments, companies and individuals can successfully address environmental problems.

Most important, working through such non-conventional analyses makes journalistic, environmental *and* economic sense: Examining whether new policy options—market incentives, for instance—could foster environmental protection at less cost to companies and individuals could become an important factor in whether such measures, and their accompanying technologies, are adopted. If considered, these factors can alter environmental debates and government policy—and inform stories. A story on local energy efficiency programs, for instance, can also examine whether state regulatory policies in some way inhibit utilities' abilities to profit from conserving energy, thereby acting as a disincentive for them to promote energy efficiency. By the same token, if a city releases a plan to cut the trash that moves into landfills, an examination of whether there are incentives for households to reduce their wastes—such as a curbside charge for the number and size of containers that each household puts out—might give a clue as to whether it will work or not.

For enterprising journalists, alternative economic views of environmental problems can also provide a mother lode of good stories. The controversial and still loosely defined notion of "sustainable development," for example, imbues a plethora of policy proposals and experiments in developing and industrial nations with economic incentives and regulatory policies, technologies and social organizations that try to lessen the environmental impact of economic development. Similarly, the fate of experiments in ecotourism, game ranching, extractive reserves and agroforestry that strive to provide a livelihood for rural people while protecting the forests and wildlife will become a factor in mitigating environmental problems in developing nations.

In the consumer societies of the developed world, the energy and water conservation efforts of cities like Los Angeles—where technological innovations such as electric cars and cleaner fuels are critical to lessening the pollution choking its skies—can also be counted among the first rudimentary experiments in sustainable development. The United Nations and several countries, such as Norway, have begun to reform the calculation of GNP to include some accounting for environmental costs and benefits. The 3-year-old International Society of Ecological Economics, a group of ecologists, economists and scientists from other disciplines, is challenging the assumptions and tools of traditional economics—everything from whether there are indeed limits to economic growth to how to develop better measures of economic welfare and calculate the value of natural resources in economic decision-making.

Covering the environment can only become more complicated as environmental hazards worsen and the economic stakes escalate. No simple prescription can ever cover the contingencies and complexities of the issues, but for journalists it will always pay to remember that in the relationship between Greens and greenbacks there is wisdom—and stories—untold in exploring alternatives.

Hallett Peak, Rocky Mountain National Park, Colorado
National Park Service

A Market for Change

Senator Timothy E. Wirth, D-Colo., is one of a new breed of policy-makers in matters of environmental protection and resource conservation. A member of the Senate Energy and Natural Resources Committee and a key supporter of the 1990 Clean Air Act, Wirth was elected to the Senate in 1986 after serving 12 years in the House of Representatives. He has been notably successful in shaping broad-based, bipartisan environmental legislation built around the idea that the old "command and control" style of regulation—a style that characterized most of the environmental legislation of the last 20 years and was embodied in the creation of the Environmental Protection Agency in 1970—is inefficient and ineffective. Instead, Wirth has proposed that economic incentives for cleanup efforts, rather than penalties for polluters, are the most cost-efficient and effective way to serve the nation's environmental goals. This approach has received widely varying reviews from the environmental lobby, with some organizations supporting the idea and others strongly opposed to it.

The idea to harness market forces in environmental policy first found a voice in Project 88, a study of market-based alternatives to current environmental programs commissioned by Wirth and his Senate colleague the late John Heinz, R-Pa., who, with Wirth, had been a driving force behind environmental policy-making in Congress. Project 88 presented the Bush administration with 36 recommendations for implementing market incentives in limiting pollution and conserving resources. Among those recommendations were imposing fees on pollution in order to encourage reduction; issuing pollution permits that corporations could buy and sell; requiring deposits on hazardous materials and granting refunds for recycled or properly disposed of waste; and ending government subsidies for timber harvesting and livestock grazing on public lands, as well as other activities that utilize public resources.

Here, responding to a series of questions from editor Craig LaMay, Senator Wirth discusses his views of media coverage of environmental issues and, in particular, the media's understanding of the economic aspects of Project 88 and environmental policy.

Given your vantage point, what do you think of the substance of the current media coverage of environmental issues? Has the media's renewed interest in this topic tended to clarify or to confuse public awareness, to inform or to sensationalize? Are there particular environmental problems you feel the public is woefully uninformed about, or are there issues where the media could do a much better job of reporting?

"The media" are not, by any means, a monolith. There is a great deal of difference between different newspapers, let alone between different media. By and large the general news media are doing more to cover the environment today than ever before. That reflects (and reinforces) the American people's interest in their own environment, and in the world's environmental problems. There are also special media which deal only with the environment, and there are cable television stations that almost exclusively run environmental documentaries.

The increase in coverage of the environment does not, however, mean that the environment is well covered in a scientific sense. The priorities the media give to various items are, by and large, far more responsive to the media's own needs than to any scientific assessment of which problems are important and which trivial.

For example, it is still true that the media do a good job of covering events and a relatively poor job of covering trends. As a result, environmental disasters like the Exxon *Valdez* spill are extremely well covered and generally well reported. But important issues like the skyrocketing growth of world population—issues that will affect the course of the planet and that call for action—are largely ignored. In a similar way, most of the improvement we have seen in environmental quality—the successes of the Clean Water Act, for example—is largely incremental, and rarely gets coverage.

Policy-makers, including those of us who are elected by the public, have a responsibility to go beyond reacting to the events of the day. In order to formulate policies that respond to the threats and opportunities facing our nation and the world, we must look at and try to understand long-term trends in order to identify policy responses that will result in success over the long term. Getting public support for such policies requires an educated and aware public.

The challenge to both policy-makers and the press is to find opportunities in the events of the day to educate the public about important developments and trends. The explosion of media coverage about global warming during the nation's hottest and driest summer was (to me) an example of the best sort of exploitation of newsworthy events: It put a complicated and complex issue concerning long-term trends before the public.

On the subject of population, however, such a convergence has not happened. Part of that is, frankly, the result of media's sensitivity to the political controversy surrounding reproductive rights in this country, and part of it is that the issue is perceived as a problem in far-off lands rather than as a challenge to the ability of humanity to maintain a livable world.

Just as important as the media's role in informing the public is their ability to capture the public's attention. Coverage of environmental issues informs the public, but perhaps even more importantly from the point of view of a policy-maker, it focuses the public's attention on these issues.

The Exxon *Valdez* spill is a good case in point. The event in itself was spectacular in its size and scope, and the coverage was both informative and sensational. But the political effects of the spill—including the passage of oil spill liability legislation, which has been hung up for the better part of the decade over issues of whether tougher state laws should be preempted—were more the result of focusing the public's attention on oil spills than of specific information about whether or not Captain Hazelwood was intoxicated.

An event such as the Exxon spill focuses the media on the subject as well. The attention given that disaster resulted in national coverage of a series of smaller spills that, had they taken place a year before the Exxon disaster, might have only gotten local coverage, if any. That helped keep the public's attention on oil spills as a problem in need of better solutions.

It is in the area of broad policy issues that the media are weakest. One of those areas has been the United States' general energy policy. There hasn't been much of a story in our growing dependence on oil imports during a period of low oil prices, though Saddam Hussein's recent actions will change that. The United States is critically dependent on foreign oil to fuel our transportation system, and we have made virtually no progress at all in the last decade in diminishing that dependence or creating alternatives to it. Another area that I have mentioned before is world population. The United States used to play a major role in this issue, but it dropped out in 1980 and its continued absence is truly tragic.

How well did the media report on Project 88? What do you say to critics who charge that your economic approach to environmental issues does not adequately address the problems created by our culture of consumption?

Considering that Project 88 was a report on concepts and environmental themes, it received quite a lot of media attention, and I was quite pleased by that. By and large, that coverage was very fair and it concentrated on the basic ideas Senator Heinz and I were trying to get across—that we can craft ways to use the strength of economic forces in our society to create more effective ways to protect the environment, and that growing environmental problems mean that we should seek out such mechanisms—even though they are very different from those used by existing environmental laws.

As for critics of Project 88, the "market-oriented" approach to environmental problems shouldn't be characterized as a "free-market" approach to the environment, or as de-regulation. It isn't either. A free market in the environment implies that people can buy environmental quality or, conversely, that polluters can buy a right to pollute that is limited only by their budget. Such a notion would be a radical change in our policy and an abandonment of the government's central role of protecting the public welfare.

What Project 88 proposed was setting up a system whereby market forces, rather than regulatory mandates, will be the driving force for reducing pollution. In the past we have told polluters how to reduce pollution, but Project 88 proposed telling polluters how much they have to reduce pollution and leaving them to decide how best to do that.

In the area of acid rain prevention, for example, what Project 88 proposed—and what President Bush and then the Congress adopted—was a system where Congress decides the overall environmental goal—in this case, a 10-million-ton reduction in sulfur dioxide emissions nationwide—and we let the polluters decide how to apportion those reductions among themselves, based largely on economic factors.

We did that by requiring power plants to have permits to cover their total pollution—but only giving out enough permits to cover about half the power plant pollution emitted today. Instead of requiring each power plant to reduce its pollution exactly in half—

which is the kind of regulation we have had in the past—we allow the plants to trade and sell their permits. A single power plant, for example, can buy permits to allow it to keep polluting at today's level, but it can only buy those permits from people who have reduced their pollution below their allotment.

Reducing pollution by half would be very expensive for some power plants and relatively inexpensive for others. Under the permit system, power plant owners will seek out the cheapest pollution reductions first, creating the most economically efficient pollution control. In effect, you get the most environmental improvement per dollar invested. In fact, the Bush administration believes that this system will achieve the 10-million-ton reduction for at least 20 percent less than more conventional approaches.

There is another advantage to the economic-incentive approach: While the uniform-prescription approach requires conformity, economic incentives reward efficiency and innovation. This sort of system is perfectly adapted to our "culture of consumption." It puts the forces of the marketplace to work to restrain pollution, and ideally, it makes it profitable to reduce pollution.

One of the problems with our current environmental laws is that they focus on detailed prescriptions for regulated parties. And as we have required higher standards of environmental protection, those prescriptions have gotten so intricately detailed that debate on environmental issues has often focused on complex and often arcane matters of technology rather than on our environmental goals. Those debates quickly leave the public—and most policy-makers—lost, having to judge between two or more camps of experts dealing in technical details.

One of the best features of the Project 88 proposal on acid rain was that it focused the debate on the goal of the program—the amount of pollution we needed to reduce. The technical details of the means to that end were left to the polluters, who were responsible for producing the results rather than for implementing a particular technology.

The usefulness of such a turn in the debate was demonstrated in this Congress. Debate was quickly narrowed down to whether a 10-million-ton reduction by the year 2000 was the correct goal, and faced with such a clear choice, Congress rather quickly decided that was a good goal. That contrasts with what had been 10 years of argument over which categories of power plants should have to install what types of pollution controls.

Do you feel that the Bush administration is setting a responsible lead in environmental matters, or is it primarily interested in image and reactive spin control?

President Bush is quite adept in his use of the media, and it is obvious that both he and his staff have made projecting concern for the environment a major priority in their media strategy. I think it fair to say, however, that their media strategy is independent of their policy agenda. On policy, the president has been good on the environment in some areas but very disappointing in most others. Of course, he emphasizes the good.

The president's primary positive on the environment has been his taking up the cause of passing Clean Air legislation. His strong endorsement of the need to pass such legislation, and his adoption of a strong, well-defined acid rain reduction program, have been key ingredients in breaking what has been a decade-long impasse on this issue in Congress.

I will say, however, that I have profound disagreements with the president and his staff on the details of that legislation and with the positions they espoused in the negotiations that led to a bill in the Senate. Many of those positions not only were in direct opposition to what I thought was best for the environment, they also represented major retrenchment from the president's public position on key issues such as whether we should require automakers to produce and sell automobiles using clean-burning alternative fuels.

Perhaps the most negative development under President Bush has been his positioning the United States as the single most reactionary force at work against the development of effective measures by the international community to address global climate change. That is a most unfortunate position for us to take, both in terms of slowing down international action on global environmental issues and in terms of discrediting the United States as a world leader on environmental protection.

There is also a lot of fluff in the president's "environmental image program"—the announcement of new programs that are little more than repackaging of existing ones, photo opportunities of the president fishing, and so on.

All of that is part of the picture the president is projecting. I believe he and his staff correctly identified the environment as something the public and the media would be giving increased attention to, and he is responding to that.

A GROWING PROBLEM for the media and the public is going to be how to judge people on environmental issues, especially as politicians at all levels are rushing to say that they are concerned about the environment and actively seeking to protect it. Doing this will require a much more critical look than it has in the past, when many politicians were quick to espouse indifference or even hostility to environmental protection. Now many of those same politicians project a picture of concern. Some have, in fact, given a higher priority to environmental protection; others have not, but don't want to antagonize a public that pollsters tell them truly does care about the environment and gives environmental protection a high priority.

Some of that critical examination must come from the editorial pages, but at the same time, reporters are going to have to spend more time seeking out critical reviews of the environmental actions, records and proposals of politicians.

Steering by the Stars

Senator Albert Gore Jr., D-Tenn., has been Congress's most knowledge-able voice on issues affecting the global environment for nearly a decade. He chaired the first congressional hearings on the issue in 1981, while a member of the House of Representatives. After 8 years in the House, Gore was elected to the Senate in 1984, and in 1988 ran for his party's nomination for the presidency. In 1989 he wrote the World Environmental Policy Act, which seeks to promote global cooperation on issues related to the greenhouse effect; and in April 1990 he chaired the first interparliamentary conference on the global environment, a program that included representatives of more than 40 nations. In addition, Gore was co-author of the 1980 Superfund Act, which created the principal federal program for cleaning hazardous waste sites and chemical spills. He chairs the Senate's Environmental and Energy Study Conference and introduced the resolution in Congress to create Earth Day 1990.

Earlier in his career, from 1971 to 1976, Gore worked as an investigative reporter for the Nashville Tennessean. Here he discusses problems he sees in environmental coverage and the role of the media in environmental policy-making.

Great Smoky Mountains National Park, Tennessee; the wilderness hosts such a variety of plant and animal life that it has been designated an International Biosphere Reserve
Tennessee Tourist Development

In your March 1989 New York Times *opinion piece, "An Ecological Kristallnacht, Listen," you close with the words of General Omar Bradley: "It is time we steered by the stars, not by the light of every passing ship." Probably no institution gets more criticism for steering by the light of every passing ship where the environment is concerned than the media. From your vantage point, do you think this criticism is justified, and if so, how could the media improve their coverage?*

Conflicts, disasters, harsh relocations—such as earthquakes, floods, drought—attract attention. When the smoke clears after a harsh battle, when the earth settles, that story recedes from public view and, therefore, from public consciousness, though the issues behind the struggle remain as critical.

Disasters spark attention from the media. Disagreements between the president and Congress, within Congress, between environmentalists and others, spark attention. And well they should. But when the disaster is over or the conflict quiets, the issues rarely receive the same attention.

For example, the Exxon *Valdez* created an environmental nightmare of historic proportions, and every night, as oil-poisoned beaches and wildlife remained, the story filled our television screens and our newspapers. When the cleanup slowed and stopped, so did the story. The issue of adequate protections, efforts to prevent another spill, and related issues such as the protection of Antarctica never received the same attention. The stories were not as compelling, the pictures not as vivid, though the issues were as important.

Or another example: the global environment. We can't see the upper ozone layer or the chlorofluorocarbons that are destroying it—allowing dangerous ultraviolet radiation to reach the earth. We can't see carbon dioxide levels in the atmosphere or sense worldwide temperature increases. We don't see the threat. It's not part of our—or the media's—daily consciousness. The political conflicts over these issues are covered far more than the issues themselves.

Consistency, persistence and a greater commitment to the issues behind the struggles and the disasters could improve coverage.

Among policy-makers, your voice has been particularly urgent and persistent on the issue of global warming. There is general agreement in the scientific community that global warming is indeed occurring, though there is some disagreement on specifics. Often, however, it seems that these narrow disagreements become grist for a journalistic presentation of the issue that makes it seem as if there is no scientific consensus at all. Fringe views are treated as if they are as meritorious as consensus views. Some journalists will argue that their professional creed of "fairness" demands such treatment, while others are now saying that it is irresponsible and a disservice to the public. What is your view of how the media portray the global-warming issue?

As a newspaper reporter at the *Tennessean* in Nashville, I learned first to always check the facts—to go look for myself to see, for example, that in fact the house had burned or that six cars were involved in the accident or that falsified records had been filed. We believe what we see, what we can hear, smell, touch. At the same time, I looked for both sides of a story, never reporting only one perspective or one argument. It was a responsibility I took very seriously.

At bottom, these are the very things impacting coverage of global climate change. Global warming defies our senses. Climate change is measured over years, not days or moments. And, for most observers who lack the advanced training of a Ph.D. scientist or the research experience of the experts, or the time to fully study the science and the evidence, it becomes merely a case of reporting both sides. On even the hottest summer day we can't say we're witnessing global warming, so if scientist A says one thing, search high and low for scientist B who will disagree. It's natural. But it creates a dangerous imbalance.

More than 700 members of the National Academy of Sciences recently wrote President Bush urging action on global climate change. Six or 7 members take the other side of the argument, but they are given equal billing with the 700. Our most eminent scientists are telling us we face a problem of unforeseen proportions.

Yet the media, compelled to search high and low for "the other side," find those who say there is no problem and amplify those voices so that soon the protests of a few are as powerful as the concerns of the many.

This is not an argument for one-sided reporting, but rather for reporting that recognizes the subtleties involved in this issue and that accurately weighs opinions and research. As importantly, it is an argument for reporting that educates the public about the issues before us.

In your view, what role should the media play in environmental affairs? Can they or should they be expected to lead the public and to prod policy-makers, i.e., to act as advocates in the environmental arena? What media organizations—whether newspapers, magazines, broadcast or cable operations—are doing the best job of covering the environment?

The media have a responsibility to inform and to educate, to tell us not only what is happening today but also why it is happening and what it will mean to us—today *and* tomorrow. They can and should not only report what is happening, but what could happen—cover not only the millions who gather to celebrate Earth Day, but also the stories we'd never see if not for their efforts—reports of thinning of the upper atmosphere, acid rain killing faraway forests. In every arena the media lead and prod policy-makers. Editorial writers and columnists, television and radio commentators, and talk shows impact policy and perception. News reporting, by drawing attention to problems and issues, moves policy. Environmental reporting is no different.

There are many in the media who are doing an excellent job. *Time* magazine, with its "Planet of the Year" edition, sparked what has been consistently strong reporting. The Cable News Network has shown a strong commitment to these issues. At the same time, there are many local and regional outlets now devoted to environmental news, such as WWOR-TV in the New York City area, that have made environmental reporting their specialty. It is difficult to say who is doing the best job, far easier to say that most could do better.

The environment as political drama.
Courtesy, TIME, Inc.

MARK O. HATFIELD

Old Growth and the Media: A Lawmaker's Perspective

In the fall of 1989, I received a call from a reporter at CBS News. The caller was interested in doing a story on the "deforestation of the Pacific Northwest."

On the surface, anybody who has heard about the spotted owl and the old-growth forests of the Pacific Northwest would not question the premise of the question. By now, we've all heard the horror stories; we've read them in the newspapers and seen the footage on the nightly news. The message is: The loggers in the Northwest have clear-cut the land and there's no more old-growth trees left, right?

Wrong.

While seemingly unassuming, the request from CBS pointed to a significant problem in the ongoing debate over the management of our national forests and, more specifically, in the debate over the northern spotted owl and our old-growth forests.

Throughout this debate, policy-makers have been plagued by several factors, some of which were directly related to the media's coverage of the issue. The problems were just as prevalent in the regional media (which one would expect to have more expertise on details) as they were in the national media.

First, there was a general lack of understanding of what is admittedly an extraordinarily complex issue. But this lack of understanding often led to inaccuracies in reporting and, in some cases, the omission of important information. When theories of forestry are presented as fact, and information provided by special interest groups is presented to the public as gospel without verification, the public is ill-served. Yet that is exactly what has happened in story after story, in broadcast after broadcast.

With the old-growth issue, policy-makers are still operating in an environment of rapidly changing information, scientific and otherwise. At the same time, we have made a supreme effort to base our decisions on the most timely and scientifically defensible information available to us. Given the fact that one of the great challenges of our age is to manage the flood of information thrown at us every hour of the day, this hasn't always been easy.

Unfortunately, we live today in a society of instant gratification. Whether it is going to a fast-food restaurant for a quick meal or buying throw-away diapers or cameras, our lives have become geared to the need for rapid response. Decisions affecting thousands of lives and the future of our planet are increasingly made with an alarming glibness.

Because of this, the responsibility of the nation's elected officials to gather, evaluate and act on rapidly changing information is a daunting one. But the responsibility doesn't stop with Congress.

I believe that the media in a free society have no less an obligation to make sure certain facts are presented accurately and fairly. While there have been exceptions, my view is that just the opposite has been true in the reporting of the "Great Forest Wars."

Additionally, of the hundreds of thousands of column inches written and miles of videotape filmed on this subject, the human dimension of this story has often been downplayed as having less importance than the biological questions—when in fact these factors are inextricably linked. In some cases, the human dimension has been ignored altogether.

The CBS request about the "deforestation of the Pacific Northwest" was a common example of the kind of misinformation being perpetuated by media accounts of the problem. Over time such statements have become part of the nomenclature for reporters and editors alike. There are other examples, such as the statement: "We are about to cut down the last of the old-growth forests." Or a variation: "Only 12 percent of the old-growth forest remains" (or 10 or 5 percent, depending on which interest group the reporter chooses to believe).

Broadcast media in particular has fallen victim to these sweeping generalizations. As any propagandist worth his salt knows, perception often becomes reality, especially if perception is repeated enough. The media, in my view, have been less-than-aggressive in challenging oft-repeated assertions. Many of these claims (the less than 12 percent old growth remaining is a good example) are repeated in the media without being checked with credentialed professionals, and the public is ill-served in the process.

Unfortunately, most facts dealing with forestry (or any area of natural science) often require considerable explanation to develop in context, requiring precious time few broadcast outlets have, and precious space newspapers no longer have. The result is the combination of film or still photos of trees crashing to earth, which inevitably skews the public's understanding of the issue. In the case of the old-growth controversy, such imagery led to thousands of calls and letters to offices of lawmakers who could only respond with clarification, fact or—more often—nothing at all.

Had reporters searching for a story on "the deforestation of the Northwest forests" researched the subject more thoroughly, they would know that the forests of Oregon and Washington cannot be compared to the Brazilian rainforests. When trees are harvested on public and private land in Oregon and Washington, new trees must be replanted in accordance with federal laws. If reforestation of a site cannot be assured, it cannot be harvested. It's as simple as that.

The result of this policy is that we now have *more* trees growing in Oregon and Washington today than there were in the 1920s. Every year Congress provides funding for the Forest Service to replant millions of seedlings and thousands of acres of forests for the next generation.

The old-growth part of the question is more complex.

There is no single definition of *old growth*, and yet reporters and editors have repeatedly (and inconsistently) used selected definitions—or parts of definitions—for *old growth* without clarification. Any statement about how much old growth remains is necessarily dependent upon which definition is used (there are at least four). To say that only 12 percent, or 10 percent or 5 percent, remains, assumes that every acre of forest land was in an old growth condition before humans logged the first tree.

Such a notion is without foundation.

Nature has, without help from humans, destroyed regenerated forests by wind, fire, insects and disease. This has, in itself, created forest lands with a variety of ages, classifications and conditions. Forests were never entirely old growth. Most forest ecologists today estimate that the forests have never been more than 20 percent old growth at any point in time. Today, using a definition developed by a Forest Service research report (PNW-447), the Forest Service estimates that there are 3.75 million acres of "old growth" in Oregon and Washington. This is 19.7 percent of the national forest land base—just about the historic average based on estimates by the forest ecologists.

The point is this: While policy-makers try to work with and make decisions on the best available information available, the media have no less an obligation to do the same. With a few exceptions in both the regional and national media, this simply has not happened.

A GOOD CASE study of this problem was a six-part "in-depth" series analyzing the national forest management controversy in the Portland *Oregonian* last year.

The state's largest newspaper of record had a golden opportunity to provide its readers with a clear and detailed analysis of the issue. When I learned the story was being written, I hoped that it would be an objective one, perhaps clarifying misconceptions, pointing out problems and errors where warranted. Instead, hundreds of thousands of *Oregonian* readers were treated to one of the longest editorials in the paper's 140-year history.

For example, the *Oregonian* repeatedly used old-growth acreage figures provided by the Wilderness Society. This definition, also valid depending on the forest characteristics being examined, varies considerably from the Forest Service definition. Although mentioning the Forest Service definition in passing, the *Oregonian* chose to use the Wilderness Society definition throughout the article to paint, in the reporter's own words, a "far bleaker" picture of the situation.

The *Oregonian*'s series, "Northwest Forests: A Day of Reckoning," went beyond objective reporting. The series took a good news story idea and turned it into a searing six-part editorial advocating a sharp reduction in forest management activities in the Northwest. In doing so, the *Oregonian* not only did a disservice to its readers, but violated the basic responsibility of the media to report "objectively."

The *Oregonian* also reported that Forest Service supervisors were "under pressure from Northwest lawmakers to meet unsustainable timber sale targets" but offered no proof of any such "pressure." I was a primary target of that accusation, and yet I can state categorically that I never pressured anyone in the Forest Service to cut more than what was sustainable.

What the *Oregonian* did not report is that Congress has actually *lowered* the volume of timber sold each year to levels below what the Forest Service considers sustainable. It is unfortunate that the *Oregonian* reporter further confused the issue by not explaining the distinction between the amount of timber *sold* annually (a level which Congress sets based on information from the Forest Service) and the amount of timber *harvested* annually (which is complicated but can be explained by analyzing the amount of timber held under contract and the requirements of the marketplace, and which *is not* determined by Congress.)

The *Oregonian* series emphasized the negative aspects of the issue and in some cases omitted key facts that would have presented a more accurate picture. Nowhere in the article was it mentioned that many of the concerns expressed about public forest management (soil erosion, watershed protection and enhancement, scenic protection, etc.) had been addressed in new forest plans now in place for each of the region's 19 national forests. The fact that 15 years and $230 million worth of work on those forest plans had led to improved forest management was incomprehensibly omitted from the series.

A basic premise of the *Oregonian* series was that Oregon is in the throes of a "painful but long predicted transition from an economy built on wholesale harvesting of its virgin forests to one more complex and diversified," but the statement is misleading. The fact is that Oregon has been steadily diversifying its economy for over 30 years. The problem is that much of the rural part of Oregon—about 70 communities and approximately 500,000 people—remains economically dependent upon forestry, agriculture and fishing. These communities do not have the flexibility to respond to sudden, unplanned change. And because 52 percent of our land base is owned by the federal government, federal decisions can create major shock waves in those towns and communities.

In addition, the *Oregonian* relied heavily on environmental sources and gave only token representation to forest industry sources in an attempt to provide an appearance of fairness. Out of a total of approximately 2,700 column inches of print, only 53 inches, or 2 percent of the total, cited industry sources, and some of them were contacted only 1 week before the series ran.

This problem of selective reporting within the context of a

news story was elevated to the national arena with a cover story in *Time* magazine, "Owl vs. Man." In what was clearly a major piece for *Time*, reporter Ted Gup spent several days in Oregon talking to industry and environmental sources to get a handle on the story. And yet, in the final analysis, Gup failed to examine in any detail what congressional solution might be offered and instead ended his story with a statement that only reflected the inadequacy of his research:

> In the case of the Northwest, the Federal Government should help retrain loggers and mill workers with grants to spur economic diversification. Congress could also help sustain the Northwest's processing mills by passing legislation aimed at reducing raw-log exports.

Mr. Gup's editorializing missed the point. First of all, Congress has provided and still is providing funds for retraining of mill workers and others displaced by the current crisis. Second, Congress has banned the export of raw logs from federal lands for over 20 years, and just last summer (before the *Time* article was written) it restored the rights of states to ban the export of logs removed from state lands.

And nowhere in the *Time* story (with the exception of a short sidebar) did we see images of displaced workers, of second- or third-generation logging families, or the schools that would close for lack of funds, or local officials who face severe budget restrictions due to the loss of timber receipts.

Instead we got brilliantly composed photographs of a logger making the final cut into a giant tree, of anti-logging protesters, of clear-cuts and of a spotted owl juxtaposed with the opening paragraph: "A lumberjack presses his snarling chain saw into the flesh of a Douglas Fir that has held its place against wind and fires, rock slide and flood, for 200 years."

As has been the case in most contemporary reporting about environmental issues, the human dimension was largely ignored in the media until only recently. Stories about shortcomings in the management of the forests by federal agencies, directives from Congress, and the "overcutting" of the forests have almost completely overshadowed the human element.

Rarely have I seen an honest perspective of the human side of this problem. And I have never seen reported the fact that counties in some of the more timber-dependent areas in Oregon receive as much as $1,535 per child from timber receipts, or that tax rates on an average home may skyrocket in some areas if a balanced solution to this problem isn't found.

For the media, it's easy to portray a quick-fix image, but the fact is, there are no simple solutions to the problem.

Certainly, we must continue to work toward true economic diversification, but we cannot define diversification as shutting off one economic faucet only to hope another will be turned on in its place. We can't define diversification as building up one sector of the economy at the expense of the other.

In the end, the media's coverage of the old-growth issue, while extensive, has been lacking in balance. Policy-makers have tried to make decisions using the best available information, and we have tried to look to the long term for reasonable solutions. We have tried to incorporate balance between the need to protect our natural resources and the need to protect a way of life for a large segment of the people whom we represent. It hasn't always been easy.

The media have no less an obligation to do the same.

THINKING
GLOBALLY

German Green demonstrators in West Berlin, 1989
Donna Binder/Impact Visuals

JOHN McCORMICK

A Green World After the Cold War?

For global politics and international relations, the past 2 years have been tumultuous. The effective collapse of Stalinism in Eastern Europe has brought about the kind of changes that were—until now—almost unthinkable. The Berlin Wall is gone, East European communist parties are being voted out of power, and there are the first glimmerings of electoral choice, real political debate and limited free enterprise in the Soviet Union.

As a political consequence of those years, Germany has been reunified, which in turn has important implications for the future of the European Community. As the 12 Community countries head toward the abolition of internal trade barriers in 1992, the attractions of membership for their neighbors apparently grow. Before the end of the century, we could well be hearing of a new Community with at least 16 members, and possibly as many as 22. No longer does the rest of the world catch cold when the Americans and Soviets sneeze; what we are seeing now may be a fundamental realigning of the global balance of power, involving very different actors than before. As Margaret Thatcher said at the Houston summit in July 1990, "There are three regional groups at this summit: one based on the dollar, one based on the yen and one based on the Deutsche mark."

But that's not all. The turn of the decade has seen another revolution that promises to have more widespread, lasting and fundamental consequences even than the end of the Cold War: the arrival of the environment as a mainstream political issue.

In the last 2 years of the '80s, the cumulative effect of an environmental movement that has now been gathering speed for more than a generation (or more than a century if you look for its real roots) has been coming home to roost. To some extent we can thank Mikhail Gorbachev. Now that we in the West can think more confidently about welcoming the Soviets as members of the global community, and talk about making cuts in nuclear weapons without the fears we had in the 1970s about Soviet advantages, we've had the time and inclination to start thinking about some of the other problems that have been jostling all these years for the attention of policy-makers and the media. The environment has long been first in line for promotion. Today, it is so high up the policy agenda that neither politicians nor editors can any longer ignore its importance as an electoral and policy issue.

Quite suddenly, it seems, world leaders are talking not only about all the usual economic, trade and defense issues, but about global warming, threats to the ozone layer, what to do about Brazilian rainforests, trade in endangered species and marine pollution. Along with trade policy and aid to the Soviets, global warming was a leading topic of discussion at the 1990 Houston summit. Television coverage of the summit was notable for the sight of correspondents having to report the technicalities of global warming with the same familiarity and grasp for detail they had previously reserved for economic and defense matters. The environment was no longer a fringe issue.

The biggest surprises have been in the ideological sphere. Because of its links with social welfare, and because environmental protection demands the kind of regulation of industry that most conservatives shun, the environment has traditionally been the preserve of liberals. But even this can no longer be taken for granted, because key anti-regulation conservatives now seem to be understanding the need for improved environmental management and protection.

Back in 1982, Margaret Thatcher referred to environmental issues as "humdrum." Just 6 years later, in October 1988, she stunned environmentalists with a speech to the Royal Society in which she argued that "stable prosperity can be achieved throughout the world provided the environment is nurtured and safeguarded. Protecting the balance of nature is one of the great challenges of the late 20th century."

Environmentalists were initially skeptical; domestically, her government had done little until then to suggest that she even vaguely understood the issue. Some even said she was using the environment to win kudos and take the international spotlight off Mikhail Gorbachev and Helmut Kohl and reassert her role as an international statesperson. But she went on to take a leading role in bringing governments together to agree to action on problems like the hole in the ozone layer and to address the apparent warming of the earth's climate. Remember, we're talking here of Margaret Thatcher, arch-free marketeer, enemy of environmental interest groups (she once described them as "subversive") and enthusiastic opponent of business regulation and increased government spending.

In the United States, meanwhile, George Bush has declared himself the environmental president. The jury is still out on this, especially as he seems to be backpedaling on global warming. Thanks largely to his caution (or perhaps that of some of his key advisers), the Houston summit ended without agreement on global warming, and without a commitment to limiting greenhouse gases. As a result, the president's performance was severely criticized by environmentalists.

On the credit side of the ledger, though, look at what George

Bush has done: taking over from a president who was just as ardently anti-regulation as Thatcher, Bush has—among other things—elevated the Environmental Protection Agency to cabinet level, committed the United States to major cuts in the sulfur dioxide and nitrogen oxides that create acid pollution and halted oil exploration off the U.S. east and west coasts until the end of the decade.

Meanwhile, at the other end of the ideological spectrum, even Mikhail Gorbachev has been calling for a concerted response to global environmental problems. Although Gorbachev's hands are obviously tied at the moment by other more pressing concerns, glasnost has revealed something that was long suspected—that socialism has left the Soviet Union and Eastern Europe with some of the worst environmental problems in the world.

What's happening here? Conservatives and socialists apparently buying the arguments of the environmental lobby? Clearly, there has been a sea change in political attitudes toward the environment as we enter the 1990s, a change so rapid and so marked that some environmentalists are beginning to worry that this may all just be a passing fad. After all, the environmental bandwagon has rolled once before. In the lead-up to Earth Day 1970 and the landmark 1972 UN Conference on the Human Environment in Stockholm, governments seemed to be falling over themselves to pass new environmental legislation and set up new environmental agencies, and environmental issues dominated the media. But the oil crises of the 1970s and the recession of the 1980s conspired to push the environment onto the sidelines of political and media attention.

Twenty years later, as we recover from Earth Day 1990, the environment is back in vogue again. But will political and media attention to the issue be just as brief and temporary as it was at the turn of the 1960s? The chances are much less likely, because circumstances are very different. There have been three particularly significant changes in the last 20 years.

First, the environment has truly been popularized. There has been a steady accumulation of scientific knowledge that has given us much more insight into the real and potential effects of environmental mismanagement. Computers, satellites and television have made us all much more aware of our place in the global environment and of the interconnectedness of cause and effect.

The most important consequence of this has been a new commitment by a growing number of businesses and industries to become environmentally responsible, and a consequent and growing consumer interest in environmentally friendly products. (Which came first is debatable.) On both sides of the Atlantic, consumers have become more used to searching their supermarkets for biodegradable paper, ozone-friendly aerosols, non-toxic household cleaners, energy-efficient light bulbs and the like. The last few months have seen a rash of new books (using up yet more trees) on how individuals can contribute to environmental protection through changes in lifestyle. The media have promoted this by giving environmental stories greater prominence than ever before and through focusing regularly on the lifestyle issue. It is almost impossible now to open a newspaper or newsmagazine, or switch on radio or television news, without hearing something about the environment or about how we can change our way of life in order to protect the environment and natural resources.

Of all the questions about the permanence of the environment as a policy issue, none is so difficult to answer as the Green consumer factor. Consumers—and marketers—are fickle and faddish, and there are already signs in Britain that Green consumerism has stopped spreading. In the summer of 1989, more than 60 percent of people classified themselves as "Green consumers" (defined as people who had sought out and bought an environmentally friendly product in the previous month). By December the percentage was still the same.

This may give some environmentalists cause for concern, but—on the other hand—it could be explained away simply by the fact that most Green consumers tend to be middle-class and that Green consumerism had already reached saturation point by summer 1989. Indeed, the percentage may decrease with time, but that is probably no cause for concern. In fact, it may be a good thing when people stop consciously thinking about the environmental consequences of consumerism and simply absorb such practices into their daily lives.

The second big difference between 1970 and 1990 is that the environment has been politicized. Yes, it is an issue with economic and scientific dimensions as well, but at the end of the day the solutions must come from governments. It is, first and foremost, a political issue, and the next presidential election in the United States will surely see the environment occupying the same kind of niche that drugs occupied in 1988. People running for office at every level are beginning to understand the importance of the environmental vote, and the late '80s have seen politicians having to familiarize themselves with environmental issues to a far greater extent than they had to in 1970. Back then the environment was still the preserve of interest groups. Interest groups still make much of the running today, but actual and would-be policy-makers are responding more thoughtfully than they did 20 years ago.

Of course it is one thing to say that something must be done to protect the environment, quite another to design and implement effective and workable policies. As a policy issue the environment is more complex than perhaps any other area of policy, including the economy. Environmental policy-makers are faced with resolving disputes between states and nations over the costs of environmental protection, with putting short-term electoral considerations aside and accepting that the environment demands the long-term view, and with somehow working out methods of budgeting for problems that mainly defy traditional benefit-cost analysis. Accepting that there is a problem is the first step—doing something effective about it is the next.

Finally, the environment has been globalized. We knew about acid rain as early as the 1850s, and global warming as early as the 1890s, but it has been only in the last 18 months that terms and phrases like *chlorofluorocarbons (CFCs)*, the *ozone layer* and the *greenhouse effect* have been filtering into everyday conversation and into the media without the need for explanatory footnotes. There is still a long way to go, and Americans still tend to be insular on these and other matters, but they now know better than ever that they can be polluted by industrial accidents in the Soviet Ukraine, that the ozone layer cannot be protected by unilateral U.S. action, and that we all have a responsibility for—and personal stake in— the removal of such apparently distant phenomena as the destruction of Brazilian rainforests or famines in Ethiopia.

Fifteen years ago international meetings on the environment were relatively rare and tended to involve only junior government representatives. Today environmental diplomacy is a prosperous growth area, and agreements on action are being reached more rapidly. For example, it took the best part of 2 decades (from the late 1960s to the late 1980s) for industrialized countries as a bloc to do something about acid rain pollution. It has taken just 5 years (1985–1990) for them to agree on action to stop damage to the ozone layer. (This relatively rapid timetable can be explained by the fact that switching from CFCs is relatively cheap, and many industries had already begun the switch in response to consumer demands, leaving governments little to do but nod in agreement.) Action on global warming may take a little longer, but the pressures are growing.

To use the favorite phrase of forecasters, if present trends continue, the 1990s are very likely to see two further changes in the nature of the environment as a policy and media issue. First, Green politics is going to become more familiar and more popular. There are still many people (probably most people in the United States) who have never heard of Green politics, and many of those who have, probably still dismiss it as a fringe concern. But consider the facts: There are active Green parties in every West European country, and their successes continue to mount. Green parties have won seats in national legislatures in Austria, Belgium, Finland, Luxembourg, the Netherlands, West Germany,

Sweden and Switzerland and in the European Parliament. West German Greens make up the fourth biggest party grouping in the West German Bundestag. Swedish Greens won 20 seats in the 1988 elections, becoming the first new party to enter the Swedish Parliament in 70 years. British Greens picked up 15 percent of the vote (but no seats) in the 1989 European Parliament elections.

There is also a rapidly emerging Green political movement in Eastern Europe and the Soviet Union, using the truly appalling state of the environment in those countries as another hammer with which to beat the nails into the coffin of central planning. East European and Soviet media are telling us all what local people have long known about the parlous state of the environment in those countries. The spate of free elections in Eastern European countries has seen new Green parties out on the hustings, winning many hearts and minds. Greens have already won 12 seats in the new 396-seat Romanian legislature, and in most other East European countries they are steadily building bases of support.

There is little doubt that—for now at least—Green ideas are becoming more popular and more familiar. Greens variously espouse a return to small-scale economics and grass roots politics, community-based solutions to society's problems, social justice and equality, self-reliance, and a new holistic view of the place of the human race on the planet. We cannot know if the Green political parties will survive; their successes in Europe have taken many of them by surprise, and their insistence on a rotating leadership is likely to prove electorally unpopular. Green ideas, however, have increasing salience, relevance and attractions in the industrialized countries that are now rejecting socialism for being too inefficient, and capitalism for being too uncaring. We may be on the threshold of a revolution in social, political and economic thinking, a Green revolution which to many is the only logical response to the excesses and internal contradictions of the agricultural and industrial revolutions.

The second likely change in the 1990s will be the continued growing prominence of environmental problems in the Third

World. As political and media attention shift away from the Cold War, and as defense monies are increasingly diverted to other sectors of the economy, so it is likely that the collective consciences of the citizens of North America and Europe will be switched to the truly appalling problems of poorer countries, and to Africa and south Asia in particular. The region faces many such problems— not least of which are the growing threat of AIDS and the persistence of civil wars and insurrections—but the 1990s will surely see further attention paid to the costs of environmental mismanagement in the tropics: deforestation, soil erosion, the spread of deserts, famine, shortages of fuelwood and urban pollution.

This is one area where the American media have a particularly critical role to play. There is little doubt that American public opinion is insular. Europeans, because of their education system, because their countries are smaller and closer together, and because of their long history as travelers and colonists, tend to be more global in their outlook. Americans, when they think about the Third World at all, tend to focus on Latin America. After all, it's closer, and the United States has long had interests in the region. But for most Americans, Africa and south Asia could just as well be on another planet. Television in particular tends to cover Africa as though it were one big country, rather than more than 50 separate states. As the relationship between East and West changes, the media must focus more on the relationship between North and South and help promote a more informed body of opinion on Third World issues.

As we enter the final decade of the century then, and as it looks increasingly as though Marxism, Leninism and Stalinism will not survive into a second century, this may be the time when we begin to focus on an entirely new set of political and economic values based on a better understanding of humanity's place on the earth. The pre-eminent fear of the postwar years has been nuclear holocaust. As that fear diminishes, we are going to have more time to look at other problems. In terms of our long-term welfare, and perhaps even our short-term welfare, the issue that demands the most attention and the quickest response is the environment.

Defoliated pine trees on the Czech-Hungarian border, 1989
Piotr Jaxa

JUDIT VESARHELYI

Hungarian Greens Were Blue

In 1988 the then Communist government of Hungary voted unanimously to stop construction of a giant dam on the River Danube. The project was already far advanced: Concrete was about to be poured into the riverbed. The government's change of position on such a long-standing project marked a significant victory for Hungary's grass roots ecological movement, better known as the Blues (from the phrase "die schoene blaue Donau"). A massive mobilization of popular opinion had rocked an apparently immovable government's insistence on proceeding with this "Stalinist mastodon," the relic of a past political era.

The nature of the political system in Hungary prevented the Blues from publicizing their concerns or lobbying the government via the mass media, and even denied to ecological activists such elementary information as the findings of government-appointed experts. In the end, however, sheer persistence won out. The Blues succeeded in creating alternate channels of communication. Petitions, *samizdat* newsletters, and appeals to the foreign press circumvented state control of the media. They also enabled Hungarian ecological protesters to make contact with their Austrian counterparts and, through them, with the world at large.

The popular stand on the destruction of the Danube River is instructive in several ways. It tells an illuminating story of the nexus between ecology and politics. Some protesters were motivated by the opportunity to vent stifled political dissatisfaction through supposedly "safer" channels; others discovered in environmental activism their only experience of the power of coordinated grass roots protest. The success of the Blues is a classic example of how informal systems of mass communication emerge when more conventional media channels are obstructed by the state. For the Hungarian Blue movement was different from parallel organizations in the West—and in no way more so than its inability to call on the resources of mass media to state its position.

THE RIVER DANUBE runs almost the length of Europe. From its source in the Black Forest it passes through parts of Germany, Austria, Hungary, Czechoslovakia and Romania to issue into the Black Sea. The Danube runs like an artery through the heart of Central and Eastern European culture. As it turns southward, it enfolds one of Hungary's most beautiful and fertile areas, whose tangible historical heritage dates back to medieval times. The river has been a powerful source of inspiration for the poets, painters and musicians of Central Europe (one need only think of Johann Strauss' "Blue Danube"), and it remains to this day an important recreational resource for its people. Clearly, any tampering with the river at any point has a dramatic effect on the ecosystems of several neighboring countries.

The possibility of taming the Danube first arose in the early 1950s. A system of hydroelectric power stations along the river would, authorities claimed, control periodic floods, thus abolishing the last obstacles to navigation. After the oil crisis of the '70s, however, the main impetus for the scheme became the need to develop alternative sources of energy. According to the plan, four-fifths of the Danube's waters were to be channeled into an artificial canal 25 kilometers long, connected to a storage lake downstream of Bratislava, Czechoslovakia. At the downstream end of the canal, in Hungary, a huge power station—the Gabci-kovo—was to be situated. Because it was to function only intermittently—providing energy only in peak periods—a second dam was to be built somewhere in the Danube Bend in Austria.

The hydroelectric project was approved in the '70s, when Hungary's political and economic reform program was stalled. The scheme was motivated primarily by political criteria: The vision of man radically reshaping his environment to utilitarian ends harmonized well with communist ideology. Virtually every individual or organization with a stake in power was an ardent supporter of the project: The party, the government, members of Parliament, politically ambitious technocrats and academics, as well as the other members of the Warsaw Pact, all jumped on the bandwagon. The project also offered the industrial lobby an enormous and prestigious investment. It was hardly surprising that environmental concerns were totally ignored.

Opposition, especially in the initial stages, was very limited. The few academics and many ordinary people who objected kept their skepticism to themselves. The crushing of the 1956 revolution had taught them the high price of resistance to the state socialist system. There were a few intellectuals who tried to withstand the state-sponsored campaign, a few experts who were convinced that the whole scheme was counterproductive. They wrote sophisticated articles on problems related to the transformation of the Danube, but the articles were published in isolated technical journals with very limited circulations.

In 1981 Janos Vargha, a Hungarian biologist and journalist, published the first article on the proposed system of hydroelectric power stations aimed at a broader public. Vargha explained that the whole project did not make sense economically, and he brought home the ecological dangers of the project to the writers, historians, artists and sociologists who read the magazine. There things remained, until 1984.

On a cold, dark evening in January 1984, Hungarian intellectuals gathered in Budapest, in a club on the banks of the Danube.

The club belonged to the Patriotic Folks' Front, a party institution but nonetheless well-known for its limited and tolerated protests against government policy. The audience was drawn by the scheduled debate between opponents and supporters of the planned construction of a dam on the Danube, and it came expecting a politically risky event that would mirror current ideas on reform. But the controversy did not materialize—not one supporter of the dam project came to the meeting. Instead, one by one, various speakers cast light on the hazards posed by the proposed construction, their views perceived by many as implicit criticism of the government.

Late that night, in an atmosphere aglow with enthusiasm, an *ad hoc* committee was formed to compose a petition. In a country where participation in "voluntary" activities was a routine civic obligation, a group of genuine volunteers was an important event. The petition was written immediately and addressed both to the Hungarian Parliament and to the government. It demanded that construction of the dam be suspended until its economic and ecological consequences could be examined.

Some people were skeptical. "Do you *really* believe that *anything* can be done against this project, even though we know that it is crazy?" they asked. But others, intellectuals and workers, city dwellers and country folk, stepped out of anonymity and signed the petition with their full names. For the first time since 1956, they had come forward. They were willing to identify themselves with the Danube River and against government policy.

One of the many problems faced by the nascent ecological opposition was finding a commonsense way of putting the issue before ordinary citizens. The water-engineering lobby used complicated terminology that no layman could understand—the implication being, of course, that such matters were best left to "experts." But the ecology lobby found itself with an embarrassment of riches when it came to raising rational and readily understandable objections. Indeed, the catalog of problems associated with the hydroelectric scheme was so long that it was difficult to know how best to approach the matter. The Blues could simply maintain that there was nothing to be gained either by the dam or by the project as a whole.

Economically the project made no sense. Apart from breaking up communities of ethnographical rarity and ruining areas of pristine natural beauty, it would also destroy large tracts of first-class arable land in Czechoslovakia. In addition, the project would endanger 25 percent of Hungary's reserves of unpolluted underground drinking water.

Nor was navigation of the Danube really an issue by this time, since river navigation had fallen well below full capacity. In any event, the river was far more likely to flood *with* the dams in place than without them, since the construction would raise the water level higher than the highest recorded flood. In Czechoslovakia, for example, the water would be *18 meters* above ground level.

Even the scheme's vaunted energy production was illusory. By the late 1990s estimated total energy production from the Nagymaros dam would be 3.6 percent of total Hungarian energy consumption. The state's own Energy Inspectorate confirmed only half of the nominal output could actually be utilized. Given that Hungary has a disproportionately high level of energy consumption in relation to its gross domestic product (GDP) (it uses up as much energy as Austria, whose GDP is five times larger), a course of rationalization, modernization and energy conservation could clearly reduce the nation's energy consumption by up to 30 percent and thus obviate the need for any hydroelectric scheme on the Danube.

In ecological terms, of course, the project was a disaster. Drastic interference of the kind involved in the hydroelectric project disrupts the functioning of the water organisms that carry out the water's natural process of self-purification. Power plants also encourage the proliferation of algae, and it takes no technical knowledge to understand that fish larvae cannot possibly withstand a wall of water several meters high rushing down the Danube twice a day. The estuaries that would have been affected are Central Europe's most intensive breeding grounds. Forests, too, would be adversely affected, and to crown it all, satellite photographs showed that both dams were to be built along geologically active fault lines!

Then there were the sociological consequences: Settlements historically tied to the river would be cut off from their raison d'etre. The whole character of the countryside would be irrevocably altered.

The pattern of events that attended the growth of the ecological opposition movement may best be described as an oscillation between official and unofficial campaigns. For a while there were continual attempts to gain official government attention, but, faced with the government's stubborn refusal to change its policy, the Blues were forced to use illegal tactics. The story of the battle to save the Danube became a race: While the Blues' campaign was gaining momentum, work proceeded as planned, and the dam came closer and closer to realization.

The Hungarian government appeared to hold all the access cards, and it was able to stack the deck by issuing inaccurate information. The estimated costs of the development, for example, were grossly underestimated. It was claimed initially the scheme would cost 12 billion forints. This escalated to 35 and later 48 billion forints. Yet given the indirect costs involved—in loss of forestry and fishing production—as well as the pollution of water resources, experts suggested that the figure was closer to 125 billion forints. Such a disparity clearly indicated a government policy of disinformation.

The opposition's strategy, by contrast, was to offer the public the necessary data and explanations to enable concerned individuals to decide for themselves. But this seemingly simple plan of action was complicated by political realities. State-controlled mass media were forbidden to cover the debate for a year following the summer of 1984. All other avenues for placing the issue on the government agenda were also blocked. The opposition's efforts to obtain official permission to establish an association, begun immediately after that momentous meeting in Budapest in 1984, were thwarted. Tentative articles of association described the goal of this organization as an effort to protect the landscape of the Danube—a perfectly innocent and, indeed, laudable undertaking—and by the end of the summer several thousand signatures had been collected. However, the move met with official suspicion and obstruction. The petition was ignored.

Faced with this official stonewalling, a dozen intellectuals decided to go underground in September 1984. They founded an organization called the Danube Circle and published a *samizdat* newsletter against the Gabcikovo-Nagymaros dam. This newsletter fulfilled the information function that in Western countries would have been carried out by the news media. Highly confidential material was published there, most of it for the first time. Thus the Hungarian Academy of Sciences was quoted describing the dam project as "a nightmare." The newsletter made debates in academic circles available to a broader public. It became a crucial tool, not only offering protesters within Hungary the sense that there really existed a channel of organized dissent, but in forging connections between the Hungarian Blues and their Green Austrian counterparts, who offered their experience and help.

From here it was but a short step to the mass media of Western Europe, and hence full circle back to Hungary. Radio Free Europe broadcast information back to millions of Hungarian listeners, thus circumventing government censorship. In 1985 the Right Livelihood Foundation awarded the Danube Circle its "Alternative Nobel Prize," citing the 10,000 signatures collected "among very unpleasant circumstances," the Circle's newsletter and the grass roots nature of the Circle's support.

By 1986 enormous popular pressure had built up on the Hungarian government to stop the construction of the dam. But much of this pressure was simply ignored. Far from addressing the issues raised by the petition, the Hungarian Parliament appointed the same commission that had drawn up the original plan to do an environmental impact study—with predictable results. A petition calling for a referendum on the hydroelectric project was similarly ignored.

But some of this political ferment began to reach official politics. In the 1985 election campaigns, although no member of the environmental opposition was actually elected to Parliament, candidates were questioned about the dam and the issues were widely discussed, largely through the environmentalists' efforts. By this stage the opposition was pursuing a two-pronged attack: One, politically safer, focused purely on environmental issues; the other took a much riskier and more overtly political line, citing macroeconomic factors.

The year 1986 saw the first joint press conference held by the Danube Circle and the Austrian Greens to protest "ecological colonization." It emerged that the Hungarian dam project was underwritten with Austrian capital—with funds available, ironically, because the Austrian Greens had prevented similar projects from being undertaken in their own country. In exchange the Hungarian government would provide Austria with free electric power for 20 years. The protesters found common course against both the democratically elected Austrian government and the non-democratic Hungarian regime.

This international protest carousel, by which Hungarian protesters inspired their Austrian colleagues to pressure their own government, and hence the Hungarian government, came about as a result of official seesaw politics in Hungary. Given economic realities, it was of great importance to Hungary to be perceived as politically liberal in the West. At the same time, however, its hard-line credentials had to "play" in Moscow. So while the contacts between the Danube Circle and the Greens abroad continued, members of the opposition were subjected to police harassment in varying degrees.

The Blue opposition was now increasingly candid. Thirty prominent Hungarian intellectuals published a full-page political advertisement in *Die Presse*, a politically neutral Viennese newspaper. A similar gesture was made by the Austrian Greens. A planned "environmental walk" inside Hungary, however, was forcibly suppressed.

But even as the opposition grew in breadth and confidence, time was running short. Construction had started. The environmental opposition found itself driven to more extreme tactics. A new, even more radical *samizdat* publication hit the streets, to be followed, in short order, by the environmental protesters themselves. Broad-based popular opposition found vivid expression in a living chain of protesters linking arms across the Danube's bridges.

By 1988, when the riverbed was prepared for the injection of concrete, protest had reached giant proportions, mobilizing a crosssection of the entire Hungarian population. Women and children went to the site, surrounded by flocks of policemen. A film was shot called *Danasaurus* (a play on "the tyrannosaurus of the Danube"). Artists published essays in a journal called *The Danube*. When what proved to be the last Communist Parliament voted almost unanimously for the construction of the dam, constituencies began to recall all but the 19 members who had voted against the project. In spring 1988, 40,000 people demonstrated in front of the Parliament buildings for 4 hours—probably the largest spontaneous public demonstration since the 1956 uprising. Faced with this undeniable, unstoppable force, Parliament voted once again: This time opposition to the construction of the dam was unanimous. Following free elections in Hungary in 1990, the restoration of the ecological unity of the Danube region became an important part of the government's program.

THE HUNGARIAN BLUE movement differs in many respects from Western movements of the same kind. For one thing, Eastern European environmental activists know that it was the state-controlled socialist apparatus which devastated the environment. Thus the Eastern European movement is much less left-wing than, for example, its Austrian and German counterparts. As well, the character of the protesters is different from those in the West.

Given the very real hardships to be faced, the people who decided to fight the Hungarian government were not seeking self-fulfillment. On the contrary, they were acting against their immediate self-interest and were motivated entirely by ethical issues. There was nothing easy about expressing environmental concern.

As the case of Hungary's Blues shows, in a system where there are few if any vehicles for popular self-expression, it becomes increasingly difficult to distinguish between environmental activism proper and political protest. Ironically enough, many parts of the Hungarian environmental movement lacked even environmental awareness! This seeming paradox is easily explained. When the state withholds access to the basic structures of democracy—the free spread of information, for example, or the right to express an opinion—any focus of popular discontent becomes a means of challenging the regime. Even if people did not fully understand the functioning of bank-filtered wells or variations in groundwater, they knew very well that they wanted to conserve their landscape, towns, forests and waterways. And they knew that the government wanted to destroy all of this.

It has become clear to all of us that no government can control all channels of communication indefinitely. Even if the official media are strictly controlled, where the opposition is great enough, unofficial channels will be created. The difference between a large-circulation, government-sponsored daily newspaper and an underground newsletter is that every word of the latter is carefully read, discussed and given credence. In the modern world of electronic media, popular discontent within a country may harness the mass media of a neighboring country to its cause, relying on public opinion abroad to influence the government at home. These are all solid lessons, painstakingly learned.

But the Blues as a political force have a long way to go in Hungary. Many former Communists saw the ecological movement as a way of salvaging a political career in changing circumstances.

There is a joke to the effect that the symbol of the Greens is the watermelon—Green on the outside, red on the inside.

In the balance sheet of the Hungarian Blue movement, we must place against the undoubted success of grass roots action—which enabled Hungarian citizens to feel, for the first time in their lives, that they could influence government decisions—a lack of focus and a tremendous, though in many ways attractive, naivete about representational politics. Despite their tremendous victory over the Communist government, the Blues are very poorly represented in the new Parliament.

We await the road ahead to see how the lessons we learned under a Communist government will serve us in the new democratic dispensation. Now it is time to take the hands of our children and see after the environment. Solid work starts now.

Victim of industrial mercury poisoning and mother: Minimata, Japan, 1971
Aileen Smith & W. Eugene Smith/Black Star

ADITIA MAN SHRESTHA

Winds from the West

When *Time* magazine made the Earth its "Planet of the Year" for 1988, it had a tremendous impact on global thinking. A similar attempt was subsequently made by an Asian magazine published in Hong Kong, but it was hardly noticed or discussed in Asia, to say nothing of other continents.

The reason for the disparity in the impact of these stories was simple: *Time*'s was a stronger story. The resources and professional research that the American magazine had invested in the preparation of its story far exceeded that of its Asian counterpart.

With the resources the Western media have at their disposal, with their awareness and initiative, they can do a superlative job on almost any environmental story they choose. On global environmental issues, particularly—the greenhouse effect, ozone depletion and the acid rain phenomenon, for instance—the West has an absolute monopoly on information. All major research efforts are taking place in the West, and all research findings are published and processed there. Only secondhand does such information reach Asia or, for that matter, Third World countries anywhere.

This has naturally given an edge to the Western media over their counterparts in the Third World. Most of the stories that appear in the Asian press originate from the West, and Asian television shows the BBC and American serials on nature conservation and science films related to environment.

It is thus not surprising to find most of the environmental stories we come across in Asia based on information supplied by the Western news agencies and media. Even some of Asia's regional and national problems cannot be discussed, let alone understood, without referring to sources of information in the West. One outstanding problem in Asia, for example, is deforestation. South Asia has the worst fuelwood crisis in the world, and logging exports have been banned in Thailand, Nepal and India. Yet deforestation cannot be reliably and quantitatively verified without the help of satellite images and monitoring records, virtually all of which are in Western hands. Statistics about forest cover, for example, vary tremendously. Nepal's forest is estimated at from 5 percent to 37 percent of its land; India's from 10 percent to 20 percent; Thailand's, 10 percent to 17 percent. Journalists are at pains to sift credibility from such a range of figures, a problem made worse by the intrusion of regional finger-pointing.

It is generally claimed, for instance, that massive deforestation in Nepal has caused unprecedented flooding in downstream countries like India and Bangladesh, such as occurred in those countries in 1988. This charge is especially frequent in the Bangladesh media. However, some scientists working at the International Center for Integrated Mountain Development, based in Kathmandu, attribute the Bangladesh floods to bad engineering there—the upsetting of the natural drainage system through the mismanagement of rivers and road building. They assert that 85 percent of the sediments carried by the big Ganges and Brahmaputra rivers is geological and that only 15 percent is caused by deforestation and other human activities, a claim which, not surprisingly, has been highlighted by the Nepalese media. The predictable result is that public opinion in Bangladesh holds Nepal responsible for the devastating floods, whereas public opinion in Nepal believes the floods are Bangladesh's own creation. Are the media misleading their people in their respective countries? Certainly they don't mean to: Discussions of deforestation, as well as other environmental issues of critical regional importance, are plagued by uncertainty even within national borders.

FROM THE PERSPECTIVE of the Asian journalist, the media in the West are overwhelmed with environmental data. The situation here is much different. Here we face technical, political and financial constraints in any reporting venture we may undertake. There is a dearth of information on environmental projects and problems, and whatever has been generated is not freely accessible but in the hands of governments or government-controlled agencies. It is either not available to journalists or filtered out to suit government interests, which are more often than not counter to public interests. What information is made available to the public is usually wrapped up in scientific jargon and technical reports, most of it incomprehensible to media people, to say nothing of the public at large.

In the big dam projects, for example, the appraisal report, impact-assessment report and evaluation report are treated by the national governments, as well as the international loaning agencies that fund them, as confidential. When contacted for such a report, they invariably refer you to the other agency. I could get hold of a copy of the impact-assessment report on the biggest hydroelectric project in Nepal, the Arun 3, only in Washington, D.C., and not in Kathmandu. That, too, was possible only on the condition that the agency supplying the report not be named.

Despite their difficulties and their shortcomings, the Asian media have raised issues that have had a direct bearing on some environmentally sensitive government decisions. Thanks to press campaigns, for instance, a controversy on radioactive milk powder, imported from Europe right after the Chernobyl nuclear disaster, became a major issue in Bangladesh, Nepal and Pakistan.

Press coverage contributed to great public pressure to stop the milk imports; however, the coverage was far from consistent. Journalists in these countries differed widely in their assessment of the significance of different levels of becquerels (an international measure of radioactivity). Local scientists, taking their cue from Western media and the World Health Organization, claimed that anywhere from 125 to 500 becquerels were safe for human consumption, and though the milk was eventually banned in all these countries, the merits of the case were never clearly spelled out.

This journalistic lack of understanding was not an isolated case. When journalists discuss projects like the big dams in the Narmada River Valley in India, the Nam Choan Dam in Thailand, the Kalabag Dam in Pakistan and the Arun 3 project in Nepal, they have a hard time explaining the intricacies of the attending environmental problems. They simply don't have an adequate knowledge base with which to approach their topic. They have similar difficulties dealing with complex topics like biodiversity, toxic waste disposal or wetland management. What often happens is that they fall back on official handouts about such matters, and the public is left confused rather than enlightened.

To plug such holes in journalists' knowledge, the Asian Forum of Environmental Journalists (AFEJ) was created in 1988 under the auspices of the UN's Economic and Social Commission for Asia and the Pacific (ESCAP). Consisting of more than 200 journalists from 11 nations—China, India, Indonesia, Malaysia, Thailand, Bangladesh, Nepal, Pakistan, Sri Lanka, the Maldives and the Philippines—the AFEJ has initiated programs that can familiarize journalists with basic scientific information with which they can better understand, or at least demystify, environmental problems. There are several information sources in Asia developed and patronized by the United Nations, agencies like the UN Environment Program, the Food and Agriculture Organization and ESCAP, as well as the regional Asian Development Bank. These agencies have pooled information on forests, livestock, fisheries, population, air quality, desertification and many other issues, and yet this information is lying idle or marginally used. There is a critical need to mobilize and encourage the Asian media to make use of this information, and the AFEJ has instituted several fellowship programs to enable journalists to take time off from their regular duties for some kind of sabbatical study and investigation. In addition, the AFEJ is attempting to organize in-depth training courses in the United States at Lehigh University, the University of Colorado, the University of Hawaii's East-West Center, and the Center for Foreign Journalists in Reston, Virginia; and another in Bangkok, at the Asian Institute of Technology. Similar programs are likely to take root at New South Wales University in Australia and the Berlin Institute of Journalism in Germany.

As well, the AFEJ is trying to develop a computer network in cooperation with the New York-based Scientists' Institute for Public Information (SIPI) that would give Asian journalists direct access to the environmental experts and resource institutions in the United States. Whether this project will work or not is yet to be seen. The AFEJ is also sponsoring regional investigative reporting efforts. A group of environmental journalists from Nepal, India and Bangladesh is scheduled to visit each of these countries, meet experts and authorities, and write joint reports for media use. Similar projects have been proposed in Southeast Asia.

For the most part, environmental issues are not nearly as politicized in Asia as they are in the West, a situation that leaves the media to function as the public's educator in environmental matters. Increasingly, however, as the public has grown more knowledgeable and aware, the media in some countries have become more activist in their environmental coverage. Journalists from India and Thailand, for instance, believe that environmental journalism is a mission rather than a profession. The Chinese media, too, take that view and claim that through their efforts a big dam project in north China was recently canceled. In the Maldives, the government, the media and the public have been preoccupied with the greenhouse effect and its accompanying sea-level rise, a phenomenon that threatens the islands' very existence. A similar concern is also apparent in the media of the coastal regions of Bangladesh, India and Indonesia, as well as the island nations of Sri Lanka and the Philippines. But the media of Third World Asia can hardly make any fresh contributions to what the world already knows about global warming and its consequences. Instead, they reiterate the fact that the developed countries are responsible for causing it and, therefore, should be responsible for controlling it. Increasingly this issue in particular is becoming an emotional rather than a rational one, a development that is hardly surprising in light of the recent resolution on this topic—or lack of it—produced by the major industrial nations at their summer economic summit in Houston.

But emotional is something that the media are not expected to be, and probably should not be. For want of substantive information or understanding, however, that is often what they are. The AFEJ's efforts are aimed at increasing journalistic and public knowledge in environmental affairs, and over the first 3 years of its existence it has taken some important steps. It has a long way to go before achieving its goal.

BOOK
REVIEW

Oak Tree, Snow Storm, Yosemite National Park, California, 1948:
Ansel Adams

ROBERT CAHN

Books (Not Thneeds) Are What Everyone Needs

A Sand County Almanac by Aldo Leopold (Oxford University Press, 1949) and *Round River: From the Journals of Aldo Leopold* (Oxford University Press, 1953).

Silent Spring by Rachel Carson (Houghton Mifflin Company, 1962).

Man and Nature by George Perkins Marsh, edited by David Lowenthal (originally published in 1864) (The Belknap Press of Harvard University Press, 1965).

Great Short Works of Henry David Thoreau, "Walking" (Harper & Row, 1982).

Small Is Beautiful: Economics as if People Mattered by E. F. Schumacher (Harper & Row, 1973).

The Lorax by Dr. Seuss (Random House, 1971).

Of all the media—radio, television, and the variety of print journalism encompassed by magazines, newspapers, newsletters and books—the case can be made that the book has had the most lasting and significant influence on environmental awareness. A purist might argue that since books do not accept advertising they are not technically a part of media as defined by Webster. And of course where breaking news is concerned, newspapers and television clearly are the chief means of communication. When stations can commit an hour of prime time to thoughtfully produced environmental "specials," television can come close to the book in communicating ideas, especially to those disinclined to read much.

But in the communication of ideas, the book remains supreme, particularly in the environmental field. It allows the space necessary for digging deeply into a subject, educating the reader in the complicated aspects of environmental issues often passed over lightly by other media.

Why have books played such an important part in the growing environmental awareness and changing attitudes of decision makers and public officials? Possibly it is due to the fact that book authors are less inhibited by the requirements of most other media for speed of delivery and a relatively unbiased chronicling of events. Authors can spend months or even years pondering the significance of ideas and let their feelings and values emerge in the book. A writer can subtly or blatantly advocate a cause and not have his or her ideas watered down by editors, publishers or network chiefs obligated to appease advertisers. Nor is the book author required to give equal space to opposing viewpoints.

Finally, the book is at hand to be read at one's own convenience. Unlike the ephemeral electronic media, the sturdy book remains available to instruct and inspire over the years, and in many cases gains in value and influence with time. This is particularly true of the books I will discuss in detail here, and the several others I will touch upon.

I THINK IT was 1968 when I was first introduced to the writings of Aldo Leopold. While researching an article, I spent some time with environmental activist Huey Johnson, who asked if I had ever read Leopold's book, *A Sand County Almanac*. Since I had not, he gave me a copy and told me how momentous the book had been in his life.

While a young fishery researcher at Lake Tahoe, California, Johnson and a crew had holed up in a cabin during a snowstorm. In idle conversation during the wait, Johnson mentioned his concern about what was happening to natural resources, and his crew chief asked him if he had ever heard of Aldo Leopold. When Johnson answered, "Aldo who?" the crew chief handed him a copy, which Johnson started reading then and there. "It changed my life," Johnson now asserts. "From then on I had a fairly clear direction as to what I wanted to do with my life." Practicing the land ethic as explained by Aldo Leopold led Johnson to a career of preserving land and natural resources. He founded The Trust for Public Land, headed the state of California's Resources Agency for 5 years, and now runs his own organization, Resource Renewal Institute, near San Francisco.

Aldo Leopold's writings also helped change my career. Just before meeting Johnson, I had started to become aware of the environmental ethic when I spent the better part of a year visiting national parks to gather information for a series of articles I was writing for the *Christian Science Monitor*. I began to see something different in the attitude visitors had toward national parks, embracing them as their own and the world's heritage, a treasure to be preserved intact for future generations, not just a playground for personal amusement. Then, when I read *A Sand County Almanac* and other Leopold writings, I began to see conservation as a way of life, a total commitment, rather than as a nice theory to be practiced occasionally when not too inconvenient. I, too, changed directions. For 3 decades my reporting and writing had ranged over such subjects as sports, the story behind the first atom bomb, the astronauts, Marilyn Monroe, and coverage of urban affairs and the Supreme Court. In 1968 I turned exclusively to environmental and conservation writing and, somewhat to my chagrin, abdicated some of my journalistic objectivity and became an environmental activist. I have never looked back, and Aldo Leopold's writings have continued to inspire me.

A professional forester whose principal interest was wildlife management, Leopold campaigned for establishment of wilderness areas within the U.S. Forest Service and was responsible for the first such unit in New Mexico's Gila National Forest in 1924. He was also a founder of The Wilderness Society and professor of wildlife management at the University of Wisconsin. While he is best known as the father of wildlife management in the United States, he was at heart a philosopher and teacher with the ability to express profound ideas in simple and entertaining language.

"Conservation is a state of harmony between men and land," Leopold wrote in his *Round River* essay "Conservation." "By land is meant all of the things on, over, or in the earth." And in his most venerated essay, "The Land Ethic," from *A Sand County Almanac*, he wrote: "The land ethic simply enlarges the boundaries of the community to include soils, waters, plants, and animals, or collectively: the land. . . . In short, a land ethic changes the role of *Homo sapiens* from conqueror of the land-community to plain member and citizen of it. It implies respect for his fellow-members, and also respect for the community as such."

Leopold died of a heart attack in March 1948, at age 61, while helping fight a brush fire on a neighbor's land near his Sauk County, Wisconsin, weekend refuge. He died knowing that *A Sand County Almanac* had been accepted for publication, but never saw a copy. (It was given final editing by one of his sons, Luna, and was published in 1949.) So incisive was Leopold's reasoning that his writings are perfectly relevant today, and will continue to be.

In "Conservation," he tells a story that provides the perfect rebuttal to logging advocates who would like us to believe that the old-growth forests will regrow and be better than ever in 60 years or so with tree-farming methods. Leopold cites Spessart Mountain in Germany, from whose south slopes came "the most magnificent oaks in the world," the ones American cabinetmakers used when they wanted the last word in quality. The north slope, which should be the better, now bears an indifferent stand of

Scotch pine, says Leopold, even though both had been managed as part of a state forest for the past 2 centuries. Why the difference? Leopold discovered that in the Middle Ages a hunting bishop had preserved the south slope as a deer forest while the north slope had been pastured and cut by settlers, then later replanted to pines. But during the period of abuse, something happened to the microscopic flora and fauna of the soil, the number of species was greatly reduced, and 2 centuries of conservation had not restored what was lost. It was simply not possible to restore the oaks.

Leopold's *Sand County Almanac* essay "Thinking Like a Mountain" provides special insight into the current issue of restoring the wolf to Yellowstone and other parts of the West. He tells how, when young and "full of trigger-itch" he was on a mountain rim-rock when he and some companions saw below them a wolf and grown pups frolicking below:

> In those days we had never heard of passing up a chance to kill a wolf. In a second we were pumping lead into the pack. . . . When our rifles were empty, the old wolf was down, and a pup was dragging a leg into impassible slide-rocks.
>
> We reached the old wolf in time to watch a fierce green fire dying in her eyes. I realized then, and have known ever since, that there was something new to me in those eyes—something known only to her and to the mountain. . . . I thought that because fewer wolves meant more deer, that no wolves would mean hunters' paradise. But after seeing the green fire die, I sensed that neither the wolf nor the mountain agreed with such a view.
>
> Since then I have lived to see state after state extirpate its wolves. I have watched the face of many a newly-wolfless mountain, and seen the south-facing slopes wrinkle with a maze of new deer trails. I have seen every edible bush and seedling browsed. . . .
>
> I now suspect that just as a deer herd lives in mortal fear of its wolves, so does a mountain live in mortal fear of its deer. And perhaps with better cause, for while a buck pulled down by wolves can be replaced in two or three years, a range pulled down by too many deer may fail of replacement in as many decades. . . . So also with cows. The cowman who cleans his range of wolves does not realize

that he is taking over the wolf's job of trimming the herd to fit the range. He has not learned to think like a mountain. Hence we have dustbowls, and rivers washing the future into the sea.

The "almanac" aspect of the book's title derives from the first section, in which Leopold leads the reader on a month-to-month tour of his beloved rural hideaway. Leopold's acute awareness of the natural world surrounding him, and his powers of expression, turned commonplace observations into delightful and educational word portraits. The writing is so entertaining that one hardly notices that Leopold is furnishing tidbits of science, ecology and, of course, social comment. In his sketches for *April*, for instance, he describes the "sky dance" of the male woodcock, which starts on the first warm evening in April at exactly 6:50. Recounting the performance as if it takes place in a theater, he remarks that the curtain goes up 1 minute later each day until June 1—the sliding time scale being "dictated by vanity, the dancer demanding a romantic light intensity of exactly 0.05 foot-candles." The male chooses as a stage a bare spot in an open woods or brush, for "the woodcock's legs are short," notes Leopold, "and his struttings cannot be executed to advantage in dense grass or weeds, nor could his lady see them there."

At the end of the essay, when the reader is thoroughly engrossed in the saga of the woodcock, the author applies his social comment: "The drama of the sky dance is enacted nightly on hundreds of farms, the owners of which sigh for entertainment but harbor the illusion that it is to be sought in theaters. They live on the land, but not by the land. The woodcock is a living refutation of the theory that the utility of a game bird is to serve as a target or to pose gracefully on a slice of toast."

Leopold constantly resisted the world's tendency to apply an economic measure to everything instead of assigning the real "value" to the land and the biotic community. "Your true modern is separated from the land by many middlemen, and by innumerable physical gadgets," he wrote in *A Sand County Almanac*.

He has no vital relation to it; . . . and if the spot does not happen to be a golf links or a "scenic" area, he is bored stiff. . . . Almost equally serious as an obstacle to a land ethic is the attitude of the farmer for whom the land is still an adversary, or a taskmaster that keeps him in slavery. . . .

The "key-log" which must be moved to release the evolutionary process for an ethic is simply this: quit thinking about decent land-use as solely an economic problem. Examine each question in terms of what is ethically and esthetically right, as well as what is economically expedient. A thing is right when it tends to preserve the integrity, stability, and beauty of the biotic community. It is wrong when it tends otherwise.

Leopold also notes in *Round River*:

We can all see profit in conservation practice, but the profit accrues to society rather than to the individual. This, of course, explains the trend, at this moment, to wish the whole job on the government.

When one considers the prodigious achievements of the profit motive in wrecking land, one hesitates to reject it as a vehicle for restoring land. I incline to believe we have overestimated the scope of the profit motive. Is it profitable for the individual to build a beautiful home? To give his children a higher education? No, it is seldom profitable, yet we do both. These are, in fact, ethical and aesthetic premises which underlie the economic system. Once accepted, economic forces tend to align the smaller details of social organization into harmony with them.

No such ethical and aesthetic premise yet exists for the condition of the land these children must live in. Our children are our signatures to the roster of history; our land is merely the place our money was made. There is as yet no social stigma in the possession of a gullied farm, a wrecked forest, or a polluted stream, provided the dividends suffice to send the youngsters to college.

Although reviewers acclaimed *A Sand County Almanac* when it was published in the fall of 1949, it was not immediately popular. Public consciousness was not ready for it until 17 years later, when

Ballantine Books put out a new paperback edition. By then the modern environmental movement was under way, and Leopold's wisdom satisfied a new hunger. Sales boomed. Millions of copies have been sold, and the book is still selling about 15,000 copies annually as a backlisted hardback.

Leopold biographer Curt Meine suggests that "a new generation of readers, eager to learn about and understand their natural surroundings, seized upon Leopold's words. . . . In books and articles, Leopold became the 'priest' and 'prophet' of the environmental movement; *A Sand County Almanac* became the movement's 'bible' or its 'scripture. . . .' As a thinker, he gave the conservation movement philosophical definition. As a poet, he enriched the nation's bookshelf of nature writing." (*Aldo Leopold: His Life and Work* by Curt Meine, University of Wisconsin Press, 1988.)

IN CONTRAST TO Leopold's gentle, introspective books on natural history, and their gradual impact on environmental consciousness, Rachel Carson's *Silent Spring* burst upon the scene surrounded by intense opposition and controversy.

A marine biologist as well as a widely read nature writer, Carson became aware that the indiscriminate use of pesticides and other poisons was posing tremendous danger to the world and its environment, then found a way to make the complex subject vividly readable and understandable. She was one of that very rare breed—a dedicated, expert scientist who had the ability to inform, entertain and motivate a reader. She has been credited with making ecology a household word, and *Silent Spring* has rightfully taken its place as a landmark literary achievement as well as a major catalyst of the modern environmental movement and the best known environmental book of all time.

"Rachel Carson designed *Silent Spring* to shock the public into action against the misuse of chemical pesticides," writes Frank Graham Jr. in *Since Silent Spring* (Houghton Mifflin Co., 1970), his assessment of the actions that took place in the years immediately following the book's publication. It "made large areas of government and the public aware for the first time of the interrelationship of all living things and the dependence of each on a healthy environment for survival" and was "the beginning of that crusade which persuaded administrators and legislators that the chemical industry would not act in the public interest unless forced to by stricter regulations."

Silent Spring is generally credited with providing impetus to the whole range of anti-pollution laws which came into force in the 1970s. Certainly the book played a part in passage of the Federal Water Quality Act of 1965 by publicizing the dangers synthetic pesticides posed to water supplies and their relation to fish kills. The public awakening generated by *Silent Spring* also helped in the long and difficult job of convincing Congress to enact new legislation to control pesticides, which it finally did in 1972. It also had a role in the eventual banning of DDT as well as the restricted use or total phasing out of all of the most notorious hard pesticides identified in the book.

Carson was known as a quiet, unpretentious scientist, bird watcher and naturalist whose books had been unpolemical works on the natural world. The literary and financial success of her best sellers *The Sea Around Us* and *The Edge of the Sea* had allowed her to leave her post at the U.S. Fish and Wildlife Service for full-time writing. It was almost by accident that she became involved in the pesticides issue in 1958. A friend sent her a copy of a letter to the editor of the *Boston Herald* complaining about the indiscriminate aerial spraying of DDT for mosquito control, which had wiped out insect life and killed many birds in a private bird sanctuary owned by the letter writer. Carson's friend wanted to know who in Washington could do something about the specific problem as well as the growing widespread use of lethal pesticides. Looking around for information to send her friend, Carson became increasingly appalled to discover that everything that meant most to her as a naturalist was being threatened.

"She was not at heart a crusader; once in a lifetime, she remarked, was enough," comments Paul Brooks, her editor at Houghton Mifflin who later wrote her biography, *The House of Life: Rachel Carson at Work* (Houghton Mifflin Co., 1972). She tried to get other writers interested in taking on the subject, even asking E. B. White to do an article for *The New Yorker*. White declined but sent her letter to the magazine's editor. When *New Yorker* editor William Shawn enthusiastically responded with an assignment, she decided to do a brief book, parts of which would appear in the magazine before publication. She put aside other writing plans and tackled the job, which she expected to finish in 4 months. Although she doubted that she could write a best seller on such a dreary theme as pesticides, she persevered because she had to, Brooks writes. "There would be no peace for me if I kept silent," she wrote to a friend at the time. The task demanded extensive research to discover the facts and especially to investigate alternatives such as use of biological controls, and it required painstaking writing and rewriting to make such a complicated subject interesting, readable and accurate. "She knew that her book must *persuade* as well as inform; it must synthesize scientific fact with the most profound sort of propaganda," says Graham. The task took 4 years and was accomplished despite Carson's increasing bouts with severe illnesses.

Any writer must stand in awe of the book that resulted. With its craftsmanship and artful weaving of science, anecdote, education and opinion, it is a unified piece of literature. From the first sentence of her brief "Fable for Tomorrow," which begins the book, the reader is drawn into the spell of Carson's writing. She describes an imaginary town in the heart of America where all had lived in harmony until a strange blight crept over the area, with farm animals, birds, vegetation and even children dying for no apparent reason. The fable concludes: "No enemy action had silenced the rebirth of new life in this stricken world. The people had done it themselves."

She points out that such a town does not actually exist, although every community has already suffered a substantial number of these disasters. Then she asks: "What has already silenced the voices of spring in countless towns in America? This book is an attempt to explain."

The enormity of the problem is established right at the start. Carson shows that the speed of change follows the impetuous and heedless pace of man rather than the deliberate pace of nature. Some 500 new chemicals are introduced each year—to which the bodies of men and animals are required somehow to adapt—and among them are basic chemicals created for use in killing insects, weeds, rodents and other "pests":

> These sprays, dusts, and aerosols are now applied almost universally to farms, gardens, forests, and homes—nonselective chemicals that have the power to kill every insect, the "good" and the "bad," to still the song of birds and the leaping of fish in the stream, to coat the leaves with a deadly film, and to linger on in soil—all this though the intended target may be only a few weeds or insects. Can anyone believe it is possible to lay down such a barrage of poisons on the surface of the earth without making it unfit for all life? They should not be called "insecticides," but "biocides."

Carson warns that since DDT was released for civilian use, the endless spiral has escalated so that ever more toxic materials must be found:

> This has happened because insects, in a triumphant vindication of Darwin's principle of the survival of the fittest, have evolved super races immune to the particular insecticide used, hence a deadlier one has always to be developed—and then a deadlier one than that. . . . Thus the chemical war is never won, and all life is caught in its violent crossfire. . . . All this is not to say there is no insect problem and no need of control. I am saying, rather, that control must be geared to realities, not to mythical situations, and that the methods employed must be such that they do not destroy us along with the insects. . . . I contend, furthermore, that we have allowed these chemicals to be used with little or no advance investigation of their effect on soil, water, wildlife, and man himself. Future generations are unlikely to condone our lack of prudent concern for the integrity of the natural world that supports all life.

After attempting to explain the complicated molecular structure of the new synthetic insecticides, whose use increased more
than fivefold between 1947 and 1960, Carson shows, chapter by
chapter, how the insecticides, herbicides and other chemicals affect soils and vegetation, surface waters and underground seas;
kill fish in polluted rivers; harm birds and all animals; and finally
act as mutagens on humankind, leading to the alarming increase
in malignant disease, especially among children. Though some of
it is difficult reading, the book serves as an ecological primer,
demonstrating the interrelationship of all things. And inter-
spersed with the ecology are chilling examples of the consequences of the arbitrary spraying of chemical biocides.

For instance, in Sheldon, Illinois, at the behest of a few farmers
and without prior consultation with U.S. or state fish and wildlife
or game management agencies, the U.S. Department of Agriculture and the Illinois Agriculture Department began a program in
1954 to exterminate the japanese beetle. Dieldrin, which was
about 50 times more poisonous than DDT, was applied at the rate
of 3 pounds per acre. The application killed not only the beetles,
but almost all insects, including earthworms; virtually wiped out
brown thrashers, meadowlarks, robins and pheasants; and killed
90 percent of all the farm cats, as well as many cattle and sheep.

The final chapter of *Silent Spring*, titled "The Other Road,"
gives the results of Carson's research into alternative, more selective ways of controlling pests with biological controls such as other
insects that feed only on the pests, or bacterial infections, or mi-
crobial insecticides. "All have this in common," she wrote: "They
are *biological* solutions, based on understanding of the living organisms they seek to control, and of the whole fabric of life to
which these organisms belong."

Even before *Silent Spring* hit the bookstores on Sept. 27, 1962,
the serialized version published in *The New Yorker* was eliciting
strident protests. It was being pilloried by the chemical industry,
the food industry, the Department of Agriculture, the farm lobby,
powerful members of Congress from farming states, and even by

some elements of the supposedly objective press. *Time* reported that "Miss Carson has taken up her pen in alarm and anger, putting literary skill second to the task of frightening and arousing readers" and inaccurately charged that scientists "fear her emotional and inaccurate outburst in *Silent Spring* may do harm by alarming the non-technical public, while doing no good for the things that she loves." *Reader's Digest* canceled a contract to print a 20,000-word condensation of the book and instead ran an abridgement of the *Time* article.

The Velsicol Corp., a leading manufacturer of two hard pesticides, chlordane and heptachlor, threatened to sue Houghton Mifflin to stop publication of the book, charging that an inaccurate statement had been made about one of its products. The publisher refused, and Velsicol took no legal action, but it continued its attacks on the book through advertising and public relations propaganda. The National Agricultural Chemicals Association spent $250,000 to discredit both the book and Carson.

She was largely vindicated when, at the request of President Kennedy (who had read *The New Yorker* articles), the President's Science Advisory Committee made a study of the values and hazards of pesticides and echoed Carson's criticism of the federal government's control programs, use of persistent pesticides and lack of concern for human safety. The scientific panel concluded that "until the publication of *Silent Spring*, . . . people were generally unaware of the toxicity of pesticides. The government should present this information to the public in a way that will make it aware of the dangers while recognizing the value of pesticides."

The controversy helped boost sales. *Silent Spring* quickly made the best-seller lists, and by December more than 100,000 copies had been sold in bookstores, with thousands more going out to the public through Book-of-the-Month-Club and other editions. Contracts were negotiated for printing 600,000 paperback copies. CBS did a TV documentary in its "CBS Reports" series on "The Silent Spring of Rachel Carson," despite the withdrawal of three of the program's five corporate sponsors. A U.S. Senate subcommittee held hearings on pesticide use, at which Carson testified. Separate editions of *Silent Spring* eventually were published in many languages, even in Chinese. She received countless honors and awards, the most important of which to her was admission to membership in the American Academy of Arts and Letters, which is limited to 50 writers, artists, sculptors and musicians.

Unfortunately, just as the opportunity came to lead the way for much-needed legislative and administrative actions to correct the problems she had revealed, the illnesses Carson had been battling for some years severely restricted her activities. She died in April 1964 at the age of 56, less than 2 years after the publication of *Silent Spring*. The book would continue to sell in bookstores and be used in schools throughout the world, selling millions of copies, to become the most-read environmental book in history and one of the top sellers for any type of book.

It is sad to note that even a powerful best-selling book was not in itself enough to overcome the more powerful political leverage of the corporate giants and their allies in Congress. Even now, more than a quarter century after *Silent Spring*, these forces have been able to delay and impede implementation of the new pesticide laws and regulations and prevent passage of the stronger legislation and enforcement that is urgently needed. The public's concern over these all-pervasive "silent poisons" was also diverted to the more readily evident issue of toxic wastes and the need to clean up the perils in the dumps "next door." Shirley Briggs, executive director of the Rachel Carson Council, a nonprofit organization formed in 1965 to advance Miss Carson's philosophy, writes in a recent magazine article ("Silent Spring: The View From 1990," *The Ecologist*, March-April 1990) that, of all the new anti-pollution laws which *Silent Spring* helped to pass, "it is ironic that the law regulating pesticides, FIFRA [the Federal Insecticide, Fungicide and Rodenticide Act], is the weakest of them all." Whereas Carson had been alarmed that production of pesticide active ingredients in the United States had risen from 124 million pounds in 1947 to 637 million pounds by 1960, the latest figures by the EPA (for 1988) show that U.S. use is now just over 1 billion pounds. "Almost none of the pesticides in use today have been adequately tested, and some not at all," Briggs told me in a recent conversation.

A CENTURY BEFORE *Silent Spring* galvanized public action, a prominent New Englander produced a book whose ideas have helped shape thinking on conservation for many generations.

George Perkins Marsh's *Man and Nature* was published in 1864, in the heyday of unrestrained natural resource exploitation. No conservation movement existed—resources were assumed to be inexhaustible, indeed, created by Deity for man's use. It was into such a setting that *Man and Nature* came with such statements as: "Man is everywhere a disturbing agent. Wherever he plants his foot, the harmonies of nature are turned to discords. . . . The Earth is fast becoming an unfit home for its noblest inhabitant." And Marsh prophesied that another era of "equal human crime and human improvidence . . . would reduce the Earth to such a condition of impoverished productiveness, of shattered surface, of climatic excess, as to threaten the depravation, barbarism, and perhaps even extinction of the species."

A lawyer and businessman, a scholar who could read in 20 languages, a teacher, congressman and diplomat, Marsh was truly a Renaissance man. President Lincoln's 1861 appointee as minister to the new Kingdom of Italy, Marsh wrote *Man and Nature* while overseas and ruminating on his observations of man's devastation of the Earth, first in his native Vermont, then in the South, then in Europe and the Middle East. He was a keen observer of everything around him and sought to find the linkages of mankind to nature, although he was not a biological scientist and thus not an ecologist in the sense in which the term is used today.

Marsh's principal concern was deforestation and its impacts on the land, yet he also noted the improvements mankind had made.

Several of the remedies he suggested—afforestation, public ownership of some natural resources, and biological controls—are a
part of today's environmental agenda. He was far in advance of his
time in urging preservation of species and their habitats, arguing
that mankind is "never justified in assuming a force to be insig-
nificant because its measure is unknown, or even because no
physical effect can now be traced to it as its origin."

On the other hand, many today who share Leopold's view of
man as part of the biotic community would disagree with Marsh's
thesis that mankind is the center of the universe and responsible
for its rise or fall. Marsh was quite comfortable with construction
of large canals or dams or various land reclamation projects, although he did warn that they should be built without harming the
environment. He was, in essence, a utilitarian conservationist.

Marsh's sentences are exceedingly wordy and pedantic, in the
fashion of his day, and it is doubtful that many except the most
dedicated have actually read the book cover to cover. Yet its
thought-provoking concepts have attracted conservationists for a
century with such ideas as: "the power most important [for man]
to cultivate, and at the same time hardest to acquire, is that of
seeing what is before him. Sight is a faculty; seeing, an art."

The book was an instant success by mid-19th-century standards, with more than 1,000 copies being sold within a few
months of publication. Before a decade had passed, according to
Marsh biographer David Lowenthal, *Man and Nature* had become
a classic of international repute. It was termed "epoch-making"
by Gifford Pinchot, another utilitarian conservationist, and was
reprinted just in time to be available for President Theodore
Roosevelt's 1908 White House Conference on Conservation. In
1931, Lewis Mumford described the book as "the fountainhead
of the conservation movement."

FAR MORE WIDELY known than Marsh is his contemporary, Henry
David Thoreau, whose literary masterpiece *Walden* has inspired
millions of readers to appreciate the natural world. His most-
quoted phrase, however, is not a part of *Walden*. "In Wildness is

the preservation of the world" comes from one of his lesser works, "Walking," which is now a part of various anthologies of Thoreau's writings.

Though "Walking" has touched modern environmental thought, it is perhaps too metaphysical for some readers. Thoreau, a protege of Ralph Waldo Emerson, was a transcendentalist (as was that other writer and towering figure of preservation, John Muir). Where George Perkins Marsh emphasized the need for observation, for seeing more than the eye sees, Thoreau went a step further to accentuate the need of being one with nature, of having a spiritual, not material or intellectual, sense of nature. "I am alarmed," Thoreau wrote, "when it happens that I have walked a mile into the woods bodily, without getting there in spirit. . . . What business have I in the woods, if I am thinking of something out of the woods?"

Thoreau professed to walk 4 hours a day and always in new fields or woods, never in roads. "Roads are made for horses and men of business. I do not travel in them much . . . because I am not in a hurry to get to any tavern or grocery or livery-stable or depot to which they lead." He claimed always to go "westward" (at least figuratively) while walking: "We go eastward to realize history and study the works of art and literature, retracing the steps of the race; we go westward as into the future, with a spirit of enterprise and adventure. . . . The West of which I speak is but another name for the Wild; and what I have been preparing to say is, that in Wildness is the preservation of the world."

Those of us who think of our profession as the "fourth estate" may be interested to know that Thoreau long ago preempted the term to describe not journalism, but the activity that he believed ranked of fourth importance in the scale of things. "The Walker," he wrote in 1862, "is a sort of fourth estate, outside of Church and State and People."

AT THE HEART of most of today's environmental battles is the inherent conflict between economics and conservation, particularly between economic growth and sustainable development. Among the many books on the controversial topic, none has been more thought-provoking or offered more environmentally sound economic solutions than E. F. "Fritz" Schumacher's *Small Is Beautiful: Economics as if People Mattered*. A British citizen born in Germany, Schumacher was a Rhodes Scholar in economics and for 20 years was the top economist and head of planning at the British Coal Board.

He was also president of the Soil Association, one of Britain's oldest organic-farming organizations, and the founder and chairman of the Intermediate Technology Development Group, which designed and built small-scale machines and devised methods of production appropriate for Third World countries. Schumacher disdained the usual ways of measuring progress, such as gross national product. He considered it foolish to judge a people's standard of living by their annual consumption, or to assume that people who consume more are necessarily better off than those who consume less. He believed that using non-renewable resources needlessly or extravagantly is an act of violence, resulting in pollution and harm to ecological systems, and threatening life itself. He warned that nations were squandering their "capital" of nature at a disastrous rate because of a misguided sense of values and wasteful, production-oriented economic systems. We treat these irreplaceable assets as income, Schumacher said, when they should be considered as our capital.

He argued that economics has forgotten what really ought to be measured, giving little or no consideration to the distinction between a dollar's worth of services on the one hand and a dollar's worth of fossil fuel or some other non-renewable resource on the other. Some "goods," such as clean air or a person's satisfaction on the job, are left entirely out of the usual economic calculations.

The solution Schumacher offered, in summary, is to get off our present collision course and begin to see the possibilities of evolving a new lifestyle designed for permanence, with new methods of production and new patterns of consumption. He suggested that a "wise" economics would base work, technology and ownership patterns on a human scale wherever possible. Schumacher did not maintain that bigness is always bad, but held that "we suffer from an almost universal idolatry of gigantism, . . . and it is necessary to insist on the virtues of smallness." The need is to use the scale that is appropriate to the activity, and Schumacher became the champion of "appropriate technology."

It may be too early to tell whether *Small Is Beautiful* (and other recent books by colleagues with revolutionary economic ideas) will eventually have enough influence to bring about the economic changes Schumacher sought. Probably not, considering government's and business's resistance to change. But if total and continuing sales are any indication of people's interest in what he had to say, there may be some stirrings of change going on beneath the surface.

In a 1977 trip that included conferences, speeches, seminars and meetings with five state governors, Schumacher spread his gospel among 34,000 people in 13 states, winding up talking to a group of 100 members of Congress and being invited to the White House for a talk with President Carter—who had read the book.

Sales of the book dropped off after Schumacher's sudden death in 1977. A new paperback edition published last year in anticipation of the 1990 Earth Day sold 11,000 copies, and more than 700,000 copies have been sold to date, making it the all-time best seller among books on economics and the environment—at least in paperback. Its popularity has surpassed that of books by colleagues with similar avant-garde ideas, such as *The Limits to Growth* by Donella and Dennis Meadows, Jorgen Randers and William W. Behrens III (Universe Books, 1972); *Creating Alternative Futures: The End of Economics* by Hazel Henderson (Berkley Publishing Corp., 1978); and *Toward a Steady-State Economy*, edited by Herman E. Daly (W. H. Freeman and Co., 1973). *For the Common Good: Redirecting the Economy Toward Community, the Environment, and a Sustainable Future* by economist Daly and philosopher John B. Cobb Jr. (Beacon Press, 1989), carries on the dialogue and presents several new ideas.

ONE UNIQUE BOOK that has had considerable impact among a vital segment of the public is *The Lorax* by Dr. Seuss, known around his La Jolla, California, home as Theodor Seuss Geisel. There is hardly an American born since 1970 (or a parent of same) who does not know the Lorax' message. In his inimitably wacky drawings and verse, Dr. Seuss tells of a Once-ler who started cutting down Truffula trees, using the soft tufts to make the all-purpose "Thneeds" ("which everyone, everyone, everyone needs"). Despite protests by the Lorax—who speaks for the trees and the birds and the fish—the Once-ler keeps "biggering" his assembly-line production, selling more Thneeds and polluting the landscape until all the creatures are forced to flee. Finally, the last Truffula tree gets axed, the factory closes, and even the Lorax departs, leaving behind a small stone monument with one word etched into it: "UNLESS." The now repentant Once-ler holes up in his tower, pondering the meaning of the mysterious word, until a small boy arrives and surveys the scene of devastation. Seeing the boy, the Once-ler at last comprehends:

> Now that *you're* here,
> the word of the Lorax seems perfectly clear.
> UNLESS someone like you
> cares a whole awful lot,
> nothing is going to get better.
> It's not.

The Once-ler gives the one last Truffula seed to the boy, admonishing him:

> Plant a new Truffula. Treat it with care.
> Give it clean water. And feed it fresh air.
> Grow a forest. Protect it from axes that hack.
> Then the Lorax
> and all of his friends
> may come back.

Karrie Fisher-LaMay

Annotated Bibliography

For a journalist, "covering the environment" means coming to grips with an extremely diverse range of issues: from local matters which pertain to a particular crisis to far-ranging blueprints for the future of the planet. As one scans the literature, one becomes aware of distinct areas of specialization: air pollution, water pollution, solid waste disposal, acid rain, the tropical rainforests, wildlife, work environments and agricultural policy; the list is endless. It is impossible to cover all aspects of environmental and ecological concern in a short note of this nature, and this bibliography cannot pretend to be an exhaustive treatment of the topic. Rather it has two precise and limited goals.

First, it provides references to some of the works by contributors to this volume, as well as readings which they suggest. For their help in this regard we would like to thank Robert Cahn, Herman Daly, Robert Gottlieb, Dana Jackson, John McCormick, Donella Meadows and Marty Strange for their valuable advice. Second, it aims to offer some suggestions to media professionals who are beginning their research into a wide range of environmental issues, and it includes most of the research and criticism in this field by journalists and journalism scholars. The bibliography is divided into five sections: "Classics," "Recent Works and Essays," "Journals and Periodicals," "Institutions and Interest Groups" and "Statistical Updating" (information published on an annual or biannual basis); where appropriate it includes short descriptions of the material.

Classics

Carson, Rachel. *Silent Spring*. Boston: Houghton Mifflin, 1962.
Commoner, Barry. *The Closing Circle: Nature, Man and Technology*. New York: Knopf, 1971.

Daly, Herman E. *Toward a Steady-State Economy*. San Francisco: W. H. Freeman, 1973.

Forrester, J. W. *Principles of Systems*. Cambridge, Mass.: Wright-Allen Press, 1968.

———. *Urban Dynamics*. Boston: MIT Press, 1969.

Giono, Jean. *The Man Who Planted Trees*. *Vogue* (1954). Repr. Chelsea, Vt.: Chelsea Green Publishing, 1985.

Glacken, Clarence J. *Traces on the Rhodian Shore: A Geographer's History of Environmental Thinking*. Berkeley, Calif.: Univ. of California Press, 1967.

Herber, Lewis. *Crisis in Our Cities*. Englewood Cliffs, N.J.: Prentice-Hall, 1968.

Leopold, Aldo. *A Sand County Almanac: With Other Essays on Conservation From Round River*. New York: Oxford Univ. Press, 1966.

Marsh, George Perkins. *Man and Nature* (edited by David Lowenthal). Cambridge: Harvard Univ. Press, 1965.

Schumacher, E. F. *Small Is Beautiful: Economics as if People Mattered*. New York: Harper & Row, 1973.

Thoreau, Henry David. *Walden*. London: W. Scott, 1886.

U.S. Government's Yearbook of Agriculture. *Grasses*. 1948.

———. *Soils*. 1936.

Recent Works and Essays

Berry, Wendell. *The Unsettling of America*. San Francisco: Sierra Club Books, 1977.

———. *Home Economics: Fourteen Essays*. San Francisco: North Point Press, 1987.

> Both Berry's books focus on the interrelation of human culture and agriculture; studies of social and economic aspects of agriculture.

———, Bruce Coleman, Wes Jackson, eds. *Meeting the Expectations of the Land: Essays in Sustainable Agriculture and Stewardship*. San Francisco: North Point Press, 1984.

Bullard, Robert D. *Dumping in Dixie: Race, Class, and Environmental Quality*. Boulder, Colo.: Westview Press, 1990.

> Examines U.S. waste disposal policies and their relation to race relations, especially in the Deep South of the United States.

Burnham, John C. *How Superstition Won and Science Lost: Popularizing*

Science and Medicine in the United States. New Brunswick, N.J.: Rutgers Univ. Press, 1987.
   Social aspects of science in the United States, and the way in which science has become incorporated into popular culture.
Cahn, Robert. *Footprints on the Planet: A Search for an Environmental Ethic.* New York: Universe Books, Incorporated, 1978.
Conservation Foundation. *State of the Environment: A View Towards the Nineties.* Washington, D.C.: Conservation Foundation/Island Press, 1987.
   An overview of past and future environmental issues, concentrating on environmental policy and protection.
Daly, Herman. *Steady-State Economics: Second Edition with New Essays.* Washington, D.C.: Island Press, 1991.
——, ed. *Economics, Ecology, Ethics: Essays Toward a Steady-State Economy.* San Francisco: W. H. Freeman, 1980.
—— and John B. Cobb Jr. *For the Common Good: Redirecting the Economy Towards Community, the Environment, and a Sustainable Future.* Boston: Beacon Press, 1989.
   Three important works which study the ethics of wealth and the environmental aspects of economic development. The author argues for a principle of equilibrium in economic terms that would be beneficial, and not harmful, to the environment.
Easterbrook, Gregg. "Everything You Know About the Environment is Wrong," *The New Republic,* April 30, 1990.
Elkington, John, Julia Hailes and Joel Makower. *The Green Consumer.* (1988.) Repr. New York: Penguin/Vintage, 1990.
   A comprehensive consumer-oriented guide, listing environmentally safe products, environmental organizations and federal offices which deal with environmental and ecological issues.
Erlich, Paul R., Anne H. Erlich and John P. Holdren. *Population, Resources, Environment.* San Francisco: W. H. Freeman, 1977.
   A study of ecoscience that relates population, environmental resources and pollution to human ecology in general.
Fox, Stephen R. *John Muir and His Legacy: The American Conservation Movement.* Boston: Little, Brown, 1981.
   A history of the conservation movement through a biography of an important 19th-century American naturalist.
Friedman, Sharon M., ed. *Reporting on the Environment: A Handbook for*

Journalists. Bangkok: United Nations Economic and Social Commission for Asia and the Pacific, 1988.

Gipps, Terry. *Breaking the Pesticide Habit: Alternatives to Twelve Hazardous Pesticides.* Minneapolis: International Alliance for Sustainable Agriculture, 1987.

Goldsmith, Edward and Nicholas Hildyard. *The Earth Report: The Essential Guide to Global Ecological Issues.* Los Angeles: Price Stern Sloan, 1988.

> *A collection that discusses man's influence on nature in articles on a wide range of issues including such subjects as the politics of food aid, nuclear energy, acid rain and so on.*

Gottlieb, Robert and Louis Blumberg. *War on Waste: Can America Win Its Battle With Garbage?* Washington, D.C.: Island Press, 1989.

> *A study of the problems of waste disposal in the United States.*

Greenberg, Michael R., ed. *Public Health and the Environment: The United States Experience.* New York: Guilford Press, 1987.

Gruchow, Paul. *The Necessity of Empty Places.* New York: St. Martin's Press, 1988.

> *A travel book which is a meditation on the American landscape.*

Hargrove, Eugene C., ed. *Beyond Spaceship Earth: Environmental Ethics and the Solar System.* New York: Sierra Club Books/Random House, 1986.

> *The environmental aspects of outer-space exploration, and an exploration of the moral and ethical aspects of environmental protection.*

Hassebrook, Chuck, and Gabriel Hedges. *Biotechnology's Bitter Harvest: Herbicide-Tolerant Crops and the Threat to Sustainable Agriculture.* Walthill, Neb.: Center for Rural Affairs, 1988.

Jackson, Wes, Wendell Berry and Bruce Coleman. *Meeting the Expectations of the Land: Essays in Sustainable Agriculture and Stewardship.* San Francisco: North Point Press, 1984.

> *Studies in agricultural policy and practice and their impact on the environment.*

Kazis, Richard and Richard Grossman. *Fear at Work: Job Blackmail, Labor and the Environment.* New York: Pilgrim Press, 1982.

> *Environmental policy and protection set off against their industrial costs, especially issues of cost-effectiveness and unemployment.*

Lovelock, James. *Ages of Gaia: Biography of Our Living Earth.* New York: Norton, 1988.

The basic book about Green politics.

McCormick, John. *Reclaiming Paradise: The Global Environment Movement.* Bloomington: Indiana Univ. Press, 1989.
   A study of citizen participation in the environmental movement; a history of environmental activism.

McKibben, Bill. *The End of Nature.* New York: Random House, 1989.
   Examines man's influence on nature and the greenhouse effect.

Miller, G. Tyler. *Living in the Environment: An Introduction to Environmental Science.* Belmont, Calif.: Wadsworth, 1990.
   A basic guide to human ecology and environmental policy.

Multinational Monitor. Several features on the General Agreements on Trade and Tariffs (GATT). Vol. 11, No. 11. November 1990.

Nader, Ralph, Ronald Brownstein and John Richard, eds. *Who's Poisoning America: Corporate Polluters and Their Victims in the Chemical Age.* New York: Sierra Club Books/Random House, 1989.
   Issues of pollution and toxicology and corporate business responsibility addressed by a leading consumer advocate.

National Research Council. *Alternative Agriculture.* Washington, D.C.: National Academy Press, 1989.

Rolston, Holmes, III. *Environmental Ethics: Duties to and Values in the Natural World.* Philadelphia: Temple Univ. Press, 1988.
   Moral and ethical aspects of human ecology.

Sandman, Peter M. *Environmental Risk and the Press: An Exploratory Assessment.* New Brunswick, N.J.: Transaction Books, 1987.

Schoenfeld, A. Clay. "The Environmental Movement as Reflected in the American Magazine," *Journalism Quarterly,* Vol. 60, 1983, 470–75.
   A review of the development of the environmental press, giving a list of journals, dates of first issues, and an overview of the growth of the media interest in the environment.

Scientists' Institute for Public Information. "Environmental Reporting," *SIPIscope,* Vol. 18, No. 1. New York, 1990.

Strange, Martin. *Family Farming: A New Economic Vision.* Lincoln, Neb.: Univ. of Nebraska Press/Institute for Food and Development Policy, 1988.
   A study of family farms, the relation between agriculture and the state, and the role of agricultural innovation.

World Association for Christian Communication. "Media and Environment," *Media Development,* London, 1990.

World Resources Institute. *The Crucial Decade: The 1990s and the Global Environmental Challenge.* Washington, D.C.: World Resources Institute, 1989.
   Environmental policy and planning for the future.
Wright, Angus. *The Death of Ramon Gonzales: The Modern Agricultural Dilemma.* Austin: Univ. of Texas Press, 1990.

Journals and Periodicals

Amicus Journal. (Natural Resources Defense Council) New York, N.Y.
Biocycle: Journal of Waste Recycling. Emmaus, Pa.
Buzzworm: The Environmental Journal. Boulder, Colo.
E* Magazine. Westport, Conn.
Ecologist: Journal of the Post-Industrial Age. Wadebridge, England.
Environment. Washington, D.C.
Environmental Ethics. Albuquerque, N.M.
EPA Journal. (Environmental Protection Agency) Washington, D.C.
Garbage: The Practical Journal for the Environment. Brooklyn, N.Y.
Journal of Environmental Education. Washington, D.C.
NAPEC Quarterly. (National Association of Professional Environmental Communicators) Chicago, Ill.
SEJournal. (Society of Environmental Journalists) Washington, D.C.
SIPIscope. (Scientists' Institute for Public Information) New York, N.Y.

Institutions and Interest Groups

Acid Rain Information Clearing House. Rochester, N.Y.
Center for Environmental Information. Rochester, N.Y.
Center for Investigative Reporting. San Francisco, Calif.
Citizen's Clearinghouse for Hazardous Waste. Arlington, Va.
Clean Water Fund. Washington, D.C.
Conservation International. Washington, D.C.
Cousteau Society.
Environmental and Energy Study Institute. Washington, D.C.
Environmental Law Institute. Washington, D.C.
Environmental Protection Agency. Washington, D.C.
Environmental Safety. Washington, D.C.
Friends of the Earth. Washington, D.C.
Greenpeace Action. Washington, D.C.

The Land Conservancy. Washington, D.C.
National Association of Professional Environmental Communicators. Chicago, Ill.
National Audubon Society. New York, N.Y.
National Clean Air Coalition. Washington, D.C.
National Wildlife Federation. Washington, D.C.
Natural Resources Defense Council. New York, N.Y.
Sierra Club. San Francisco, Calif.
Society of Environmental Journalists. Washington, D.C.
The Trust for Public Land. San Francisco, Calif.
U.S. Department of the Interior. Washington, D.C.
U.S. Council on Environmental Quality. Washington, D.C.
World Wildlife Fund. Washington, D.C.

Statistical Updating (Annual or Biannual Editions)

Brown, Lester, et al. *State of the World*. New York: W. W. Norton.
Directory of Environmental Investing. Available from Environmental Economics, Philadelphia, Pa.
National Wildlife Federation. *Conservation Directory*. Washington, D.C.
Population Reference Bureau. "World Population Data Sheet." Washington, D.C.
Sivard, Ruth Leger. *World Military and Social Expenditures*. Washington, D.C: World Priorities.
UNICEF. *The State of the World's Children*. New York: Oxford Univ. Press.
World Bank. *World Development Report*. New York: Oxford Univ. Press.
World Resources Institute. *World Resources*. New York: Oxford Univ. Press.
Worldwatch Institute. *State of the World*. New York: Norton, 1991.

List of Contributors

John Maxwell Hamilton, a former foreign correspondent and senior World Bank public affairs official, is author most recently of *Entangling Alliances: How the Third World Shapes Our Lives*.

Sharon M. Friedman is chair of the department of journalism at Lehigh University, where she directs the science and environmental writing program. She is co-author of *Reporting on the Environment: A Handbook for Journalists* and a fellow of the American Association for the Advancement of Science.

John Burnham is professor of history at Ohio State University and author of *How Superstition Won and Science Lost: Popularizing Science and Medicine in the United States*.

Robert Gottlieb is coordinator of the environmental analysis and policy area of the UCLA urban planning program and co-author of *War on Waste: Can America Win Its Battle With Garbage?* and *Thirst for Growth: Water Agencies as Hidden Government in California*.

Everette E. Dennis is executive director of The Freedom Forum Media Studies Center at Columbia University and a vice president of The Freedom Forum in Washington, D.C. He is the author or editor of several books on mass media.

Donella H. Meadows is coordinator of the International Network of Resource Information Centers, an adjunct professor of environmental studies at Dartmouth College, and a syndicated columnist.

Teya Ryan is senior producer and co-developer of Turner Broadcasting's "Network Earth" and a vice president of the Society of Environmental Journalists.

Jim Detjen has reported about science and environmental issues at the *Philadelphia Inquirer* since 1982. He is a five-time winner of the Scripps Howard Foundation's national award for environmental reporting and serves as the president of the Society of Environmental Journalists.

Craig L. LaMay is editor of the *Gannett Center Journal*, a former reporter and medical science editor. He is a member of the Radio and Television News Directors Association's Advisory Council on Environmental Reporting.

William J. Coughlin is the former editor of the *Washington Daily News*. He now teaches journalism at Francis Marion College in Florence, South Carolina.

Gerry Stover is the former executive director of the Environmental Consortium for Minority Outreach in Washington, D.C., a non-profit membership organization working to increase the participation of minorities in the environmental movement.

Dana Jackson is director of education and editor of *The Land Report* at the Land Institute in Salina, Kansas.

Herman E. Daly is a senior economist at the World Bank. He is the author of *Steady-State Economics* and co-author of *For the Common Good*.

Emily T. Smith is the science editor for *Business Week*.

Timothy E. Wirth is a Democratic U.S. senator from Colorado.

Albert Gore Jr. is a Democratic U.S. senator from Tennessee.

Mark O. Hatfield is a Republican U.S. senator from Oregon.

John McCormick is visiting lecturer in political science at Indiana University-Purdue University at Indianapolis and author of *Reclaiming Paradise*, a history of the global environmental movement.

Judit Vesarhelyi is a writer with the Independent Ecological Center, a program of the Soros Foundation in Budapest, Hungary.

Aditia Man Shrestha is chairman of the Asian Forum of Environmental Journalists and lives in Kathmandu, Nepal.

Robert Cahn won the 1969 Pulitzer Prize for national reporting for a series of articles on the national parks for the *Christian Science Monitor*, and is the author of several books on the environment. He was one of the original members of the President's Council on Environmental Quality and presently serves on the boards of Friends of the Earth and The Trust for Public Land.

Index

"ABC Evening News," 84
"ABC Nightly News," 4
acid rain, 175–177, 201
Adams, Michael, 123
advertising
  environmental, 130
  political, 107
  TV commercials, 149–151
advocacy reporting, 58–59
alternatives to, 100–101
arguments in favor of, 81–89, 93–94
criticism of, 93–94, 121–122
debatable as a term, 81
effects of, on media credibility, 165
vs. objectivity, 26, 60–61, 103–113
affirmative action, 127, 129, 131
Africa, environmental crisis in, 203
agribusiness, 140–141, 144–146
agricultural plants, origins of, 5–6
agriculture
  chemical use, 136–140
  in developing countries, 8, 10, 143
  environmentalists and, 141–142
  in Europe, 145
  harmful practices, 135–142
  high-yield imperative, 139–140
  as industry, 140–142
  organic, 137
  sustainable (alternative), 141–142, 163
  traditional, 140–142, 146

air pollution, 29–31, 161, 175–177
Alar scare, 24–25, 109, 137
Alliance for a Clean Rural Environment (ACRE), 138
Alternative Agriculture (NAS report), 138
American culture
  paradigms of, 73–74
  TV commercials and, 149–151
American public
  environmental awareness, 87, 232
  view of environmental movement, 31
  view of nature, 60
Amstutz, Daniel, 143
apocalyptics, environmental, 37
Asia
  environmental coverage in, 25–26, 217–222
  environmental crisis in, 203, 218–222
Asian Forum of Environmental Journalists (AFEJ), 26, 220–222
Audience Research and Development, 18
audiences
  for environmental coverage, 83–84, 109
  mainstream and "converted," 83

media, size of, 105–106
Auletta, Ken, 107
Austria, 212

background reporting, in environmental coverage, 20–21, 24–27,
96–98, 181–182
balanced reporting, 82, 105. *See also*
objective reporting
pros and cons of, 84–87, 89
Bangladesh, 218–219
BASF, campaign against, 51
Berry, Wendell, 70
Bhopal accident, 23
biological diversity, 3–12
preserving, 11–12
biological pest controls, 236
biology science writing, 34–36
biopolitics, 29–41
Blues (in Hungary), 205–215
books
about environmental issues, 225–
244
value of, vs. other media, 225–226
Briggs, Shirley, 238
Brooks, Paul, 234
Brower, David, 144
Brown, Lester, 86, 93
Bukro, Casey, 96
Bullard, Bob, 125
bull commercials, Merrill Lynch,
149–151
Bush administration, 142, 163, 175–
178, 197–198
business, opposition to, 62
BusinessWeek, 18
Butz, Earl, 136
Buzzworm, 19

Cable News Network (CNN), 17,
83–84, 183
California, environmental ballot
propositions, 138, 164–165
carcinogens, 117, 136–137

Carson, Rachel, 232–239
Silent Spring, 29, 36, 37, 138–139,
232–239
Carter administration, 243
cash crops, in developing countries,
8, 10
"CBS News," 185
"CBS Reports," 237
Center for Media and Public Affairs,
19, 92
Charlotte Observer, 115
Chavis, Benjamin, 128
chemical industry, 137–139, 142,
144–146, 236–239
chemicals
agricultural, 136–140
newly synthesized, 235
Chernobyl accident, 23
chlorofluorocarbons (CFCs), 159,
161
Citizens for a Better Environment, 52
Clean Air Act of 1990, 161, 166
Clean Water Act, 91
climate change, 86, 182–183, 197
Codex Alimentarius, 143–144
Cold War, end of, 195–196
collective action, vs. individual action,
47, 87, 155
Committees on Occupational Safety
and Health (COSH), 50
Commoner, Barry, 54
Communists, 203, 214–215
competition, as societal value, 142
CompuServe, 83–84
computer data bases, in investigative
reporting, 100
conflicts of interest, of media, 115–
116
consensus (Lippmann), 108
conservation movement, 45
ethic of, 228–232
precursors of, 239–240
conservation programs (e.g., soil),
144

conservatism, social, media and, 74–77

Conserving the World's Biological Diversity (report), 7

consumerism, 46–47, 53, 70

consumers, Green awareness among, 199–200

control systems, 68–71

controversy, news value of, 40, 164, 182–183

Corcovado National Park (Costa Rica), 11

Costa Rica, 3–12

cost-benefit analysis, 166–168

credibility, of media, 94, 165

crisis reporting, impact on background environmental coverage, 20–21, 24–25, 96–98, 181–182

Cronkite, Walter, 98

culture, common, 71–72
  decline of, 107

Cutlip, Scott, 108

Danube Circle, 211

Danube dam project, opposition to, 205–213

Darwin, Charles, 8

debt-for-nature swaps, 11

deer, 229–230

deforestation, 228–229, 239
  in Asia, 218
  in Pacific Northwest, 185–192

democracy, public debate and, 106–107

Detjen, Jim, 25

developing countries
  agriculture, 8, 10, 143
  environmental crisis in, 202–203
  foreign debt, 10–11, 163
  and industrialized countries, 11–12
  nature parks, 11

Dewey, John, 112

Donora, Pennsylvania, smog crisis, 29–31

drugs, medicinal, origins of, 5

Dumanoski, Diane, 94

dust storms, 135–136

E magazine, 19

Earth Day
  1970, 31, 44–46, 53, 198
  1990, 39, 53, 157, 198

Easterbrook, Gregg, 13

Eastern Europe
  Green Politics in, 202, 205–215
  political changes in, 195

ecology. *See* environment

economic analysis, 166–168

Economic and Social Commission for Asia and the Pacific (ESCAP), 220–221

economic growth
  vs. economic development, 153–155
  ideology of, 150–155
  traditional measures of, 8–10, 70, 152–154, 166–168, 242
  uncounted costs of, 152–154

economics
  fallacies of, 151–155
  macro vs. micro, 154
  people-centered, 242–243

economy
  optimal scale of, 153–154, 241–243
  steady-state vs. growing, 153
  U.S., problems in, 150–151

editorials, 103–104

Efron, Edith, 37

Egolf, Brenda, 23

Emerson, Ralph Waldo, 72, 241

"empowerment," 104

energy policy, 48, 174

environment
  economic value of, 166–168, 242
  public awareness of, 87, 232

Environmental Action, 44

environmental activism, individual and grassroots, 47, 87

environmental books, 225–244

environmental coverage. *See also* jour-
    nalists, environmental; media
advocacy. *See* advocacy reporting
amount of, 17–19, 92–93, 95–96,
    172
audience for, 83–84, 109
awards for, 17, 115
background reporting in, 24–27,
    96–98, 181–182
beginnings of, 36–38
boring nature of, 55, 60
critics of, 20–21
as current hot topic, 4, 13, 18–19
difficulties of, for journalists, 4, 6–9,
    13–14, 20–22, 162–169
environmentalists' unhappiness
    with, 61–62
faults and deficiencies of, 36–38,
    52–54, 96–98, 185–192
in foreign countries, 25–26, 217–
    222
issues and topics reported, 22–24,
    47–48, 55–64, 88, 131–133, 158–
    159
objective. *See* objective reporting
and policy-making, 39–40, 183
priorities in (news values), 55–64
in the public interest, 18–19, 84–85
style and methodology of, 24–27,
    36–41, 55, 60, 96–98, 164, 181–
    183
successes and degree of progress of,
    19–28, 92–94, 172–174
suggestions for improving, 95–101
training needed for, 14
environmental crisis
in foreign countries, 198, 202–203,
    218–222
as global issue, 98–99, 196–203
industry and, 29–31, 157–169
measurement of, 9, 97–98
minorities as affected by, 127–128,
    131–132
as national security issue, 12–13

overpopulation and, 150–151
politicization of, 200
popular awareness of, 198–201
solutions to, 100–101. *See also* envi-
    ronmental protection
Environmental Defense Fund, 45
environmental groups, grass-roots,
    43–54
media coverage of, 47–48
monitoring of pollution, 87, 91–92
and NIMBY, 49–50
environmental groups, national
conflicts with grass-roots groups,
    43–54
and minorities, 125–133
professionalization of, 45–46, 50
salary levels, 129–130
staffing of (minorities underrepre-
    sented), 127–131
Environmental Index (proposed),
    97–98
environmentalism, "lifestyle," 52–53
environmental justice and democracy,
    51–53
environmental movement
and agricultural interests, 141–142
beginnings of, 31, 34–37, 44–45
class, race and gender conflicts in,
    47, 49–52
labor unions in, 50–51
minorities in, 50, 125–133
motives and tendencies of (anti-
    business, anti-scientific), 35–36,
    62
national vs. grass-roots groups, 43–
    54, 125–133
popularization of, 36–37
public's view of, 31
recent trends (radicalization), 52–54
relations with media, 61–64
women in, 49–50
environmental protection
economic costs of (job loss, etc.),
    160–162, 165–167, 191–192

global action, 161
individual action for, 47, 155, 199–200
industry's contribution of expertise to, 159–160
market-oriented approach to (incentives), 175–177
as political issue, 200
regulatory approach to, 196
Environmental Protection Agency, 22, 96, 171, 198
regulations, 118–119, 139
Europe
agriculture, 145
Green politics in, 201–202, 205–215
European Community, 195
Executive Trend Watch, 92
exploitation of resources, 8–10
Exxon *Valdez* oil spill, 24, 174

fairness. *See* balanced reporting; objective reporting
Farm Bill of 1990, 141–142
farming. *See* agriculture
farm policy (e.g., supports), 139–146, 163
Federal Insecticide, Fungicide and Rodenticide Act (FIFRA), 139, 238
feedback, information as, 68–71
fertilizers, chemical, 136, 139
food
processing companies, 145–146
safety regulations, 142–144
trade policies, 142–144
foreign countries
environmental coverage in, 25–26, 217–222
environmental crisis in, 198, 203, 218–222
forestry, 76, 144, 185–192, 228–229

Fortune magazine, 159
free-market ideology, 142
Fuller, Buckminster, 79

Gámez, Rodrigo, 5, 10, 12
Garbage: A Practical Journal for the Environment, 19
GASP (Pittsburgh), 44
Geisel, Theodor Seuss, 244
gene banks, 11–12
General Agreements on Trade and Tariffs (GATT), 142–146
genetic technology, 6
Gerbner, George, 39
Get Oil Out (Santa Barbara), 44
global environmental issues, 196–203
awareness of, 54
coverage of, 98–99
interdependence of, 99, 161
U.S. policies affecting, 163–164, 178
global warming, 86, 182–183
Goodland, Robert, 9
Gorbachev, Mikhail, 198
Gore, Sen. Albert, Jr., 179
Gorney, Carole, 23
government
officials, as news sources, 21–22
policy, economic effects of, 167–168
Graham, Frank, Jr., 232
grass-roots. *See* environmental groups, grass-roots
Gray, Betty, 123
Great Plains, 135–136
green consumer movement, 157–160, 199–200
Greenpeace, 52
Green politics, 201–202, 213
in Europe, 201–202, 205–215
gross national product, as measure of economic progress, 9, 70, 154, 166
Group of Ten, 125–133
Gup, Ted, 191

Hair, Jay D., 125
Hammond, Allan, 103–104
Harper's, 19
Harris, W. Franklin, 11
Harris-Cronin, John, 91
Hayden, Tom, 164
Hays, Samuel P., 35
health
  public, 37
  worker, 50–51
Heckadon, Stanley, 85
Hertsgaard, Mark, 7, 13, 26, 93
Hook, Sidney, 111
Houston summit, 196–197
Hungary, 205–215
hybrid crops, 6
hydroelectric projects, 206–213,
  219–220

India, 218
individual environmental action
  vs. collective action, 47, 87, 155
  consumer awareness and, 199–200
  personalized ("life-style"), 155
industry
  and environmental crisis, 29–31,
    157–169
  environmental stance of, 157–160
  media coverage of, 88
  as news source, 35–36, 190–191
  as special interest group, 78
information
  as feedback, 68–71
  useful and useless, 82, 106, 111–
    112
informationsphere, 67–68, 79
Intermediate Technology Development Group, 242
international agreements on the environment, 161, 201
international politics, after the Cold
  War, 195–196
International Society of Ecological
  Economics, 168

Investigative Reporters & Editors,
  100
investigative reporting (environmental), 99–100
  suggested guidelines for, 121–124

Jackson, Jesse, 125
Javna, John, *50 Simple Things You Can
  Do to Save the Earth*, 19
Johnson, Huey, 226–227
Jordan, David Starr, 32
journalism
  ethics of, 81–82, 88–89, 94, 103–
    105, 111, 121–122
  future of, 107, 112–113
  and the public interest, 18–19, 62–
    63, 84–87, 106–107
journalists (in general)
  cultural authority of (to teach), 109–
    110
  status of, 58
  writing style of, 103
journalists, environmental. *See also*
  science writers
  background of, in public affairs, 36
  criticisms of, 20–21
  as educators, 27, 109–111, 152
  as environmental advocates, 13
  numbers of, 95–96
  training of, 92, 96

Kennedy administration, 237
KOB-TV, 18
KSL-TV, 18
Kuhn, Thomas, *The Structure of Sci-
  entific Revolutions*, 71

labor unions, in environmental move-
  ment, 50–51
land, value of, 230–231
Lasch, Christopher, 82, 106–107
La Selva Biological Station, 3, 5
Lee, Charles, 128
Leopold, Aldo, 224–232

Round River, 231
A Sand County Almanac, 224–232
"life-style" environmentalism, 52–53
Lippmann, Walter, 104, 107, 108, 112
lobbying, environmental, 45–46
Los Angeles, 168
Los Angeles Times, 18, 84, 159, 165
Lowenthal, David, 240

MacNeil, Robert, 109
market approach to environmental protection, 107, 142, 175–177
Marsh, George Perkins, *Man and Nature*, 239–240
Matthews, Jessica Tuchman, 89
Meadows, Donella H., *The Limits to Growth*, 70
media
audience size, 105–106
books in relation to, 225–226
conservative nature of, 74–77
credibility of, objectivity and, 94, 165
criticisms of, 75–77
educational role of, 106–107, 109–111, 131–133, 152, 173, 183
environmental coverage. *See* environmental coverage
nature and paradigms of, 38–39, 74–77
relations with environmental movement, 61–64
relations with social institutions, 63–64, 74–77
Media Monitor, 19, 24
Media Resource Service, 22
medicinal drugs, origins of, 5
Meine, Curt, 232
Mendez, Fidel (Costa Rican farmer), 7–8
Merrill Lynch, bull commercials, 149–151
Mexico, 163

minorities
effects of environmental problems on, 127–128, 131–132
in environmental movement, 50, 125–133
growth of, nationally, 127
Mitchell, Wesley C., 33
Montreal Protocol, 161
Muir, John, 241
Mumford, Lewis, 240
Murrow, Edward R., 88
Muskie, Edmund, 47
Myers, Norman, 12

National Academy of Sciences, 138, 182
National Agricultural Chemicals Association, 237
National Coalition Against the Misuse of Pesticides, 137
National Public Radio, 17
National Resources Defense Council, 24, 45, 109, 136–137
National Wildlife Federation, 45
nature
Americans' view of, 60
future of, 77–78
nature parks, in developing countries, 11
neighborhood groups. *See* environmental groups, grass-roots
Nelkin, Dorothy, 35, 40
neo-imperialism, 11–12
Nepal, 218–219
"Network Earth" (Turner Broadcasting), 83–86
network news, 84
New Age Journal, 132
news
controversy as, 40, 164, 182–183
environmental topics as, 55–64
news-worthiness criteria, 55–58, 62–64, 172–174
perennial topics, 57

sources of (press releases, etc.), 21–22, 35–36, 108, 190–191
newspapers
  environmental coverage by, 18
  local, problems of, 115–116, 121–124
  shrinking audience for, 106
news system, faults of, 105–109
Newsweek, 19
New York magazine, 19
New York Times, 18, 159
NIMBY (not in my backyard) politics, 49–50
Nixon, Richard, 46
Norway, 168
nuclear power, 48

objective reporting
  vs. advocacy, 26, 60–61, 103–113
  criticisms of, 81–82, 93–94, 108–109
  as journalism ethic, 59
  support for, 94
objectivity
  misuse of, 138
  in science, 33
oil, 12
Oil, Chemical and Atomic Workers, 51
old growth forest, 188
Oregonian (Portland), 189–190
organic farming, 137
O'Rourke, P. J., 88
overpopulation. *See* population crisis
owl vs. logging story, 76, 185–192
ozone layer depletion, 98, 201

Pacific Northwest, deforestation, 185–192
Panama, 85–86
paradigms, 71–77
  of American culture, 73–74
people-centered economics, 242–243
People's Pipewatch, 91

Persian Gulf crisis, 12
pesticides, 136–140, 143, 232–239
Pimentel, David, 140
Pinchot, Gifford, 240
"Planet of the Year" story (*Time* magazine), 4, 17, 85, 183, 217
plants. *See* agricultural plants
policy-making, environmental
  media coverage and, 39–40, 183
  as political issue, 200
politicians, environmental record of, 178
politics
  environmental, 200
  Green, 201–202, 205–215
  market research in, 107
pollution
  citizens' monitoring of, 87, 91–92
  control incentives ("permits"), 171–172, 175–177
  control legislation, 46–47, 158–159
  global, 163–164
  measurement of, 97–98
  public's fear of, 87
population crisis, 150–151, 173, 174
Port Huron Statement (SDS), 44
poverty, 151–152, 154–155
Poverty and the Environment, 132
press releases, 107–108, 138
progress. *See also* economic growth
  traditional measures of, 8
Project 88, 171–172, 174–177
propaganda, media and, 75
P3, The Earth-Based Magazine for Kids, 19
public, the. *See* American public
public affairs, lessened interest in, 106–107
Public Broadcasting System (PBS), 18, 76–77, 82–83
public health, 37
public interest, journalism and, 18–19, 62–63, 84–87, 106–107

public relations, and the media, 107–
108, 138
public service reporting, vs. advocacy
reporting, 100–101
Pulitzer prizes, for environmental
coverage, 17
Pyle, Barbara, 93

Rachel Carson Council, 238
radioactive contamination, 219–220
Rapetto, Roberto L., 160, 166–167
Reader's Digest, 237
Reagan administration, 107
reform movements, environmental-
ism's roots in, 44–45
regulatory approach to environmental
protection, 196
Resource Renewal Institute, 227
resources, exploitation of, 8–10
Riddle, Oscar, 34
Right Livelihood Foundation, 211
risk analysis reporting, 24–25, 27
Robinson, Joan, 152
Rocky Mountain News, 18
Ruffins, Paul, 132
Russell, Cristine, 27

Safe Drinking Water Act of 1979,
118, 122
Schoenfeld, Clay, 36
Schudson, Michael, 33
Schumacher, E. F., 73, 242–243
Small Is Beautiful, 242–243
science
anti-scientific attitudes, 38–39
controversy in (media attention to),
40, 182–183
funding sources of (private), 110–
111
popularization of, 31–36
scientific method, 32–33, 40–41
science writers. *See also* journalists,
environmental
deficiencies of, 33–36, 38–41

scientific writing vs. science writing
(by journalists), 103
and scientists, rivalry with, 38–41
sources for, 22
Scientists' Institute for Public Information (SIPI), 18, 22, 92, 221
Seattle Times, 99
Seuss, Dr. (Theodor Seuss Geisel),
244
Shabecoff, Philip, 125, 127, 132
Shistar, Terry, 137
Sibbison, Jim, 22
Sierra Club, 45
Sigma Delta Chi, 99
Smithsonian magazine, 19
social justice issues, 51–53, 130–131
society, values of, and consensus,
108–109
Society of Environmental Journalists,
58, 100
Soil Conservation Service, 136
soil erosion, 135–136
sources of environmental coverage
industry vs. citizens groups, 35–36,
190–191
press releases, 107–108, 138
types of (breakdown), 21–22
use of information technology, 26–
27
Soviet Union
environmental problems in, 198
Green politics, 202
political changes in, 195–196
species loss, 4–12, 150, 152–153
spheres of the world (litho-, hydro-,
atmo-, bio-, information-), 67–68
Sports Illustrated, 19
survival alliances, 13
systems analysis (field), 67–71

technology, opposition to, 35–36
television commercials, and American values, 149–151
television news

and newspaper news (cooperation), 122

shrinking audience for, 106

Thatcher, Margaret, 196, 197

"Thinking Like a Mountain" (Leopold essay), 229–230

Third World. *See* developing countries

thneeds, 244

Thoreau, Henry David, *Walden*, 240–241

Thorwarth, Alfred, 99

Thurm, Scott, 100

Tichenor, Phillip J., 34

Time magazine, 19, 191, 237
"Planet of the Year" story, 4, 17, 85, 183, 217

toxics
campaigns against, 51–52
waste disposal, 48–49

trade policy
foodstuffs, 142–144
and global pollution, 163–164

tropics, biological diversity of, 3–12

Trust for Public Land, The, 227

Tuck Industries, 91–92

tuna boycott, 87

Turner Broadcasting System, 83, 93

TV Guide, 19

United Nations, 25, 168
agencies, 220

United Nations Conference on the Human Environment (1972), 198

United States
economy, problems with, 150–151
policies affecting the global enviroment, 163–164, 178
settlement of, environmental effects of, 8, 239

United States Department of Agriculture, 139–140

United States Forest Service, 189, 228

Urban Underground (New York City), 44

U.S. News and World Report, 19

Utne Reader, 132

Van Sloten, Dick, 9

Vargha, Janos, 207

Velsicol Corp., 237

Voss, Mike, 123

Wald, Matthew L., 13

Wall Street, ideology of, 150–151

Wall Street Journal, 159

Ward, Morris "Bud," 26

Washington, North Carolina, 117–124

Washington (N.C.) *Daily News*, 99, 115–124

Washington Post, 159

waste disposal, 48–49

water pollution, 115–121

Water Quality Act of 1965, 233

White House Conference on Conservation (1908), 240

Wilderness Society, 45, 189, 228

wildlife, value of, 228–230

wildness (Thoreau), 241

Wilkins, Lee, 23

Winship, Tom, 95, 100

Wirth, Sen. Timothy E., 171

Witt, William, 20

wolves, 229–230

women, in environmental movement, 49–50

Woolard, Edgar S., Jr., 158

worker health, 50–51

Worldwatch Institute, 98

writing style, popular vs. specialist, 103

WWOR-TV, 183

Yeutter, Clayton, 142

yuppies, 130

Also Available from Island Press

Ancient Forests of the Pacific Northwest
By Elliott A. Norse

Balancing on the Brink of Extinction: The Endangered Species Act and Lessons for the Future
Edited by Kathryn A. Kohm

Better Trout Habitat: A Guide to Stream Restoration and Management
By Christopher J. Hunter

Beyond 40 Percent: Record-Setting Recycling and Composting Programs
The Institute for Local Self-Reliance

The Challenge of Global Warming
Edited by Dean Edwin Abrahamson

Coastal Alert: Ecosystems, Energy, and Offshore Oil Drilling
By Dwight Holing

The Complete Guide to Environmental Careers
The CEIP Fund

Economics of Protected Areas
By John A. Dixon and Paul B. Sherman

Environmental Agenda for the Future
Edited by Robert Cahn

Environmental Disputes: Community Involvement in Conflict Resolution
By James E. Crowfoot and Julia M. Wondolleck

Forests and Forestry in China: Changing Patterns of Resource Development
By S. D. Richardson

The Global Citizen
By Donella Meadows

Hazardous Waste From Small Quantity Generators
By Seymour I. Schwartz and Wendy B. Pratt

Holistic Resource Management Workbook
By Allan Savory

In Praise of Nature
Edited and with essays by Stephanie Mills

The Living Ocean: Understanding and Protecting Marine Biodiversity
By Boyce Thorne-Miller and John G. Catena

Natural Resources for the 21st Century
Edited by R. Neil Sampson and Dwight Hair

The New York Environment Book
By Eric A. Goldstein and Mark A. Izeman

Overtapped Oasis: Reform or Revolution for Western Water
By Marc Reisner and Sarah Bates

Permaculture: A Practical Guide for a Sustainable Future
By Bill Mollison

Plastics: America's Packaging Dilemma
By Nancy Wolf and Ellen Feldman

The Poisoned Well: New Strategies for Groundwater Protection
Edited by Eric Jorgensen

For a complete catalog of Island Press publications, please write:
Island Press, Box 7, Covelo, CA 95428, or call: 1-800-828-1302